Restaurant Franchising

Restaurant Franchising

Second Edition

Mahmood A. Khan

JOHN WILEY & SONS, INC.

New York • Chichester • Weinheim • Brisbane
Singapore • Toronto

This book is printed on acid-free paper. ∞

Copyright © 1999 by John Wiley & Sons, Inc. All rights reserved.

Published simultaneously in Canada.

No part of this publication may be reproduced, stored in a retrieval system or transmitted in any form or by any means, electronic, mechanical, photocopying, recording, scanning or otherwise, except as permitted under Sections 107 or 108 of the 1976 United States Copyright Act, without either the prior written permission of the Publisher, or authorization through payment of the appropriate per-copy fee to the Copyright Clearance Center, 222 Rosewood Drive, Danvers, MA 01923, (978) 750-8400, fax (978) 750-4744. Requests to the Publisher for permission should be addressed to the Permissions Department, John Wiley & Sons, Inc., 605 Third Avenue, New York, NY 10158-0012, (212) 850-6011, fax (212) 850-6008, E-Mail: PERMREQ @ WILEY.COM.

This publication is designed to provide accurate and authoritative information in regard to the subject matter covered. It is sold with the understanding that the publisher is not engaged in rendering professional services. If professional advice or other expert assistance is required, the services of a competent professional person should be sought.

Library of Congress Cataloging-in-Publication Data:

Khan, Mahmood A.
    Restaurant franchising / by Mahmood A. Khan. — 2nd ed.
      p.   cm.
    ISBN 0-471-29194-3 (cloth : alk. paper)
    1. Restaurant management.   2. Franchises (Retail trade)
I. Title.
TX911.3.M27K49 1999
647.95'068—dc21   98-30334

Printed in the United States of America.

10 9 8 7 6 5 4 3 2

To my family—Maryam, Samala, Feras, and Nufayl—for their support, patience, and affection

CONTENTS

Preface vii

Chapter 1 Introduction to Franchising 1

Chapter 2 Restaurant Franchising in the U.S. Economy 21

Chapter 3 Pros and Cons of Franchising 63

Chapter 4 Franchising Agreements and Legal Documents 89

Chapter 5 Franchisee/Franchisor/Franchise Selection 119

Chapter 6 Franchise Application and Franchise Package 145

Chapter 7 Standard Franchisor Services 181

Chapter 8 Financial Aspects of Franchising 213

Chapter 9 Franchisor-Franchisee Relationships 243

Chapter 10 International Franchising 269

Chapter 11 Franchise Concept Development 315

Chapter 12 Nontraditional Franchises 361

Index 393

PREFACE

As we reach the new millennium, franchising will play a greater role in any industry, and more so in the restaurant business. The second edition of *Restaurant Franchising* is the direct result of the success of the first edition. Several revisions were made in response to the changing business environment, based on the comments received from numerous domestic and international readers. This edition is also designed to meet the needs of the academic and industry professionals. *Restaurant Franchising,* second edition, starts with an introduction of what franchising is and is followed by an illustration of how much of an impact it has on the U.S. economy. Also described in detail are the advantages and disadvantages of franchising, from both the franchisors's and the franchisee's point of view. An effort is made to explain some of the legal documents that are required by franchisors and franchisees.

Selecting a franchise, franchisee, and franchisor is discussed from the point of view of their desired attributes and qualifications. This is followed by an assessment of the contents of a franchise package and the application process. Standard services provided by the franchisor are covered also. All these aspects are essential for understanding the intricate nature of franchising.

Franchising demands a unique symbiotic relationship between the franchisor and franchisee. As much as this relationship is mutually beneficial, it is very fragile. Misconceptions and misunderstandings may cause a strained relationship between the franchisor and franchisee. Critical points in this relationship are fully explored and explained. This topic becomes equally important, since the U.S. Congress has been examining this relationship.

The last two chapters in the book have been completely rewritten in light of the recent and forthcoming changes. Franchising has crossed boundaries and reached worldwide, and it has a tremendous potential for further international development. Points to consider in international franchising and related factors are addressed in a separate chapter.

Some readers of the first edition have been entrepreneurs and people who would like to start their own franchise business. In order to address their needs, the chapter on franchise concepts was revised to illustrate the stepwise approach to a franchise concept development.

A new chapter was added to this edition, which deals with nontraditional franchises. Since these franchises are becoming extremely popular, an in-depth discussion is included. Charts, tables, graphs, and other illustrations are added to highlight the critical points of the discussion.

With all of its changes, *Restaurant Franchising* continues to be the only book of its kind in the domestic and international markets. Several academic programs not only adopted the first edition of the book but also introduced franchising as a new course within their curriculum. This edition will further enhance their educational and research pursuits.

Several new descriptions of franchise restaurants have been added that can be used as case studies to supplement information included within the pages of this book. Each description has been expanded and uniquely focuses on different aspects from historical to contemporary business practices.

It is the intent of the author to make this edition as useful as possible to franchisors, franchisees, prospective franchisors and franchisees, students, and all those who are interested in the area of franchising. Much effort went into collecting and revising materials from different sources, without which it would have not been possible to complete a work of such magnitude.

The author wishes to acknowledge the support and advice of colleagues and friends who helped to complete this edition. The author would especially like to acknowledge the cooperation of the following organizations and restaurant franchise corporations that have been extremely generous in providing and permitting the use of information included in this book: The National Restaurant Association; The International Franchise Association; Nation's Restaurant News; Wendy's International, Inc.; A&W Restaurants, Inc.; Golden Corral Steaks, Buffet & Bakery, Inc.; Long John Silver's Restaurants, Inc.; Godfather's Pizza, Inc.; Arby's, Inc.; AFC Enterprises; Pizzeria Uno Corporation; Captain D's Seafood Restaurants; Subway Restaurants; Made in Japan Teriyaki Experience; Burger King Corporation; International Dairy Queen, Inc.; Domino's Pizza, Inc.; Big Apple Bagels, Inc.; Blimpie International Corporation; McDonald's Corporation; and Hardee's Restaurants. In addition to these are several other restaurant corporations that willingly provided information that is included in this edition of the book.

The author wishes to thank all those who helped in reviewing, copy editing, and publishing this manuscript, and hopes that this book will serve as a tool for success for all readers who are in pursuit of knowledge pertaining to restaurant franchising. The final product, which is based on several years of work, would not have been possible without the patience, affection, and understanding of my wife and children. The author is indebted to John Wiley & Sons Editors Claire Thompson Zuckerman, JoAnna Turtletaub, and Pamela Chirls for their cooperation, advice, and help during the development and completion of this project. Also, many thanks to Donna Conte for her help during the final stages of the production of this book.

CHAPTER 1

Introduction to Franchising

Franchising—the most dynamic business arrangement—has become the dominant force in the distribution of goods and services in the United States as well as in many other parts of the world. National and international experts have predicted that franchising will become the primary method of doing business worldwide. Paradoxically, in spite of its popularity and distinct impact on the economy, franchising still remains a relatively obscure concept. Some view it as an industry in itself, while others associate it with a particular type of business, such as fast-food restaurants. Most of the confusion stems from franchising being an umbrella term that covers a wide variety of business arrangements and activities. It is not restricted to a particular type of business activity. In fact, its strength lies in its adaptability to an ever-expanding array of industries, markets, and products, in addition to being responsive to economic development and the consumer demands. Simply stated, franchising is essentially a *method* of distributing goods and services.

FRANCHISING DEFINED

The term *franchise* originated from a French word meaning "free from servitude." Strictly from the business point of view, a franchise is a right or privilege granted to an individual or a group. Franchises may be granted by government or private bodies. From the point of view of economics, a franchise is a right granted to operate a business under the general regulation of one who grants it. Simply defined, a franchise is a legal agreement in which an owner (franchisor) agrees to grant rights or privileges (license) to someone else (franchisee) to sell the product(s) or services under specific conditions. This method of doing business is referred to as *franchising* and, like marketing or distributing a product or service, may be adopted and used in a wide variety of industries and businesses.

From a legal standpoint, according to a report by the Committee on Small Business, United States Congress (1990), "franchising is essentially a contractual method for marketing and distributing goods and services of a company (franchisor) through a dedicated or restricted network of distributors (franchisees)." Under the terms of this legal franchise contract, a franchisor grants the right and license to franchisees to market a product or service, or both, using the trademark and/or the business system developed by the franchisor. The contract imposes obligation on both parties. The franchisor must provide the product, a proven marketing plan or business format, management, and marketing support and training. The franchisee brings financing, management skills, and a determination to own and operate a successful business. According to B.R. Smith and T.L. West (1986), "a franchise is a legal agreement between two parties wherein each party gives up some legal rights to gain some others." In the best arrangement everyone wins—the franchisor expands its number of outlets and gains additional income; the franchisee has a business of his or her own.

According to L.T. Tarbutton (1986), a generally accepted definition for the term *franchise* is "a long term, continuing business relationship wherein for a consideration, the franchisor grants to the franchisee a licensed right, subject to agreed-upon requirements and restrictions, to conduct business utilizing the trade and/or service marks of the franchisor and also provides to the franchisee advice and assistance in organizing, merchandising, and managing the business conducted pursuant to the licenses."

R. Justis and R. Judd (1989) define franchising as "a business opportunity by which the owner (producer or distributor) of a service or a trade-marked product grants exclusive rights to an individual for the local distribution and/or sale of the service or product, and in return receives a payment or royalty and conformance to quality standards." Although similar to the above definitions, this one takes into account the conformity to quality standards by a franchisee.

The U.S. Department of Commerce (in Kostecka, *Franchising in the Economy, 1985–1987*) defines franchising as a method of doing business by which a franchisee is granted the right to engage in offering, selling, or distributing goods or services under a marketing format which is designed by the franchisor. The franchisor permits the franchisee to use the franchisor's trademark, name and advertising."

The International Franchise Association, the major franchising trade association, defines franchising as "a continuing relationship in which the franchisor provides a licensed privilege to do business, plus assistance in organizing, training, merchandising, and management in return for a consideration from the franchisee." In a widely circulated publication entitled *Investigate before Investing,* the association offers a further definition of *franchising* based on a prototype franchise disclosure law: "a contract or agreement either expressed or implied, whether oral or written, between two or more persons by which: (a) a franchisee is granted the right to engage in the business of offering, selling or distributing goods or services under a *marketing plan or system prescribed in substantial part by a franchisor;* and (b) the operation of the franchisee's business pursuant to such plan or system is *substantially associated with the franchisor's trademark, service mark, trade name, logotype, advertising or other commercial symbol designating the franchisor or its affiliate.*"

BASIC CONCEPTS OF FRANCHISING

From the above definitions certain concepts become distinctly apparent and can be summarized as follows:

- *Franchising* is a method of distributing goods and services.
- A *franchise* is a right or privilege granted to an individual or a group.
- A *franchise* is a legal agreement between two parties.
- The owner who agrees to grant rights or privileges is referred to as the *franchisor.*
- The individual or group to whom the rights or privileges are granted by the franchisor is called the *franchisee.*
- The system under which franchisor and franchisee operate is known as *franchising.*

The above definitions are crucial in understanding the concept of franchising. A franchise should not be confused with a subsidiary or branch operation of a business. A business may have several solely owned subsidiary branches of the original operation that are not franchises. For example, Sears® has branch stores that cannot be considered as franchises. The use of trademark alone does not constitute a franchise. There are multi-unit operations and groups of restaurants owned by an individual or corporation that have the same trademark but are not franchises.

For example, Red Lobster® restaurants are a part of a national chain; individual restaurants are not franchises. What constitutes a franchise is the legal agreement between a franchisor and a franchisee for the conduct of specific business. Further, a franchise-granting corporation may itself be a wholly owned subsidiary of another corporation. A good example is Pizza Hut®, which is a subsidiary of Tricon Global Restaurants®, which also owns Taco Bell® Corporation and Kentucky Fried Chicken®. Thus Pizza Hut is a component of a large conglomerate. Components of conglomerates are not considered franchises, although some of them may individually be franchise-granting corporations.

In summary, under the terms of the franchise contract, a franchisor grants the right and license to franchisees to market a product or service, or both, using the trademark and/or the business system developed by the franchisor.

The entire process of franchising starts with a *concept,* which may be based on an idea, name, process, product, or format. The franchisor grants a license to another party to use this concept. The franchisor normally charges a fee for this arrangement, which is called a *franchise fee.*

FRANCHISING: A SYMBIOTIC RELATIONSHIP

A review of restaurant franchises reveals the existence of a symbiotic and mutually beneficial relationship between the franchisor and the franchisee. If properly executed, franchising is a win-win-win situation. There are significant advantages to franchisor, franchisee, and the consumer. For an entrepreneur, a small business, or a growing company with a potentially successful product, process, or plan, franchising provides a cost-effective and systematic strategy for marketing and rapid expansion with a minimum of direct involvement and financial investment. For a prospective franchisee, it represents an opportunity to own and operate a business involving a proven concept, product, or business format with a minimum of financial risk. For potential consumers, franchising provides a way to receive goods and services in a reliable and predictable manner.

For the franchisee, the most significant characteristic of a franchise relationship is the minimizing of the risk of starting a new business. A franchisee also benefits from consumer recognition of the franchisor's trademark and products. Costly operating and marketing mistakes can be avoided because franchisors provide advertising, training, continuous supervision, and assistance. On the other hand, a franchisee is not as independent as a non-franchised businessperson because of his or her contractual dependence on the franchisor for promotion, advertisement, training, technical support, maintenance of quality standards, and overall assistance in operational matters. This loss of independence is often the cause of friction and conflict. Other franchisor-franchisee relationship problems can be traced to the franchisee's yielding of some options and controls. The franchisor exercises a certain degree of control over the actions of the franchisees, primarily for the maintenance of quality and performance standards. The degree of control varies from one franchise to another and is based on the type of business.

Types of Franchise Arrangements

A variety of business arrangements may exist within a franchise system. All of these arrangements can be classified into two major groups, as described below.

Product and Trade Name Franchising

Product and trade name franchising began primarily as an independent sales relationship between supplier and dealer in which the dealer acquired some of the supplier's identity. The dealer (franchisee) identifies with the supplier (franchisor) through the product line and, to some extent, with its trade name or trademark. Franchisees are granted the rights to distribute a franchisor's products within a specified territory or at a specific location, generally with the use of the manufacturer's identifying name or trademark. Examples of product and trade name franchising include automobile dealerships, gasoline service stations, soft-drink bottlers, and farm equipment dealers. This type of franchising dominates the field. There is considerable competition and, to a certain extent, saturation among franchises within this category.

Business-Format Franchising

Business-format franchising involves a complete business format rather than a single product or trademark. It is a relatively new kind of franchising and is characterized by an ongoing business relationship between franchisor and franchisee. Business-format franchising includes not only product, service, and trademark but also the entire business concept itself—a marketing strategy and plan, operating manuals and standards, quality control, group purchasing power, research and development, and a continuous process of training, assistance, and guidance. The franchisee is required to comply with the franchisor's guidelines pertaining to all aspects of the business, including operating procedures, the quality of the products and/or services, and the physical appearance of the business facility. A two-way channel of communication is maintained.

Examples of business-format franchising include restaurants (all types); hotels, motels, and campgrounds; recreation, entertainment, and travel; automotive products and services; business aids and services; construction, home improvements, maintenance, and cleaning services; convenience stores; laundry and dry cleaning; educational products and services; rental services (auto and truck); rental services (equipment); nonfood retailing;

and food retailing (nonconvenience). Business-format franchising is responsible for much of the franchising in the United States and internationally since 1950. It has shown rapid growth and continues to offer numerous opportunities for individuals seeking to own a business.

Differences between the two types of franchising sometimes appear blurred, particularly among nonrestaurant franchises. Where alternative business arrangements, such as distributorships and licensed agencies, are concerned, it becomes hard to place franchises in one group or the other. In some cases, it is a question as to whether to classify them as a franchise or not. Given the absence of a single, generally accepted definition of a franchise, the Committee on Small Business, 101st U.S. Congress, in a 1990 report, states that the most commonly recognized criteria for determining whether a business arrangement is a franchise is the one incorporated by the Federal Trade Commission (FTC) in its 1978 regulatory rule governing disclosure requirements for franchises and business opportunities ventures (FTC Rule 436). Under this franchise rule, a continuing commercial relationship is classified as a franchise if three characteristics are present:

1. The franchisee distributes products or services associated with the franchisor's trademark or identifying symbol.
2. The franchisor provides significant assistance and/or exercises significant control over the franchisee's method of operation.
3. The franchisee is required to pay at least $500 to the franchisor during the first six months of operation of the franchise.

All states incorporate criteria similar to the FTC's franchise rule, although their definitions may differ, particularly with regard to the threshold amount paid to acquire a franchise. Thus, regardless of what a business arrangement is called, how it is advertised or sold, or even whether it has a formal written contract or agreement, the contractual relationship between parties in the business arrangement governs whether or not it is a franchise.

The dynamic nature and the rapid development of franchising will certainly change the nature and shape of future classification of franchising arrangements and will lead to the introduction of new business terminology. New and creative methods of business arrangements (creative franchising) are being introduced to keep pace with the growth of franchising. Some increasingly common terms are *dual-concept franchising, master franchising,* and *conversion franchising.*

Dual-concept franchising refers to the arrangement in which two concepts function simultaneously at one location. For example, some gas stations are teaming up with restaurants or baked goods franchises and providing both types of goods and services at one location. This concept lends itself to effective cost-sharing and increase in sales by providing additional goods and services to captive customers. Customers stopping at a gas station may find it convenient to buy from a franchised restaurant or doughnut shop located at the same premises.

The master franchising program is designed to provide growth of a franchise in areas or locations where original franchisor may not have easy access. In this program, an individual is trained to be a franchisor and, upon completion of a set training, is referred to as a *master franchisor.* A master franchisor is responsible for selling a franchise and for assisting the new franchisee with the total franchise package, including the site selection, equipment purchase, and personnel training. The master franchisor is also responsible for maintaining the quality of product and/or services in the assigned region on an ongoing basis. The master franchisor assumes all functions of the original franchisor and acts as sole representative in the region. This type of arrangement works efficiently in international markets, where the original franchisor may not have easy access due to political, social, or cultural reasons.

In conversion franchising, an owner of an existing business decides to become a franchisee by associating with a franchisor.

For example, a restaurant owner may decide to convert his or her existing business into a nationally recognized franchise. The franchisor benefits by established clientele and tested business site, while the franchisee benefits by recognition and the franchisor's assistance. Conversion franchising can prove financially beneficial for both parties.

HISTORY AND DEVELOPMENT

Franchising in principle, if not in its current form, has existed for many centuries. In early ages, kings and rulers granted the right to certain individuals to collect taxes. In the Roman republic, officials referred to as *publicani* were responsible for collecting taxes, a portion of which they withheld as compensation. In the medieval period, churches granted individuals privileges to conduct business enterprises within their jurisdiction. In England, many companies received charters of incorporation from the crown.

The rapid development of franchising started in the late 1800s, coinciding with the Industrial Revolution. Changes became evident in the way business was conducted and innovative distribution methods were sought. All these industrial and business changes, coupled with the mass movement of populations to cities and suburbs, led to the development of franchising. Individual enterprises found it profitable to expand into larger franchises, particularly in real estate, hardware, auto repair, and other retail businesses.

The first formal form of franchising in consumer goods firms was in the year 1851, when Isaac Singer accepted fees from independent salesmen to acquire territorial rights to sell his recently invented sewing machine. Singer Sewing Machines® had experienced difficulty in marketing the innovative new product; there was a need for representatives to go out and educate consumers regarding its versatility. Because Singer did not have the capital to hire such a large workforce, agents working on commission

Introduction to Franchising 11

were the most logical choice. Franchising gained much broader recognition with the marketing efforts planned by the automobile industry. General Motors Corporation sold its first franchise in 1898, after which franchising became common throughout the automobile and gasoline industries.

Franchising in the Early 1900s

The success of franchising in the auto and petroleum industries opened the door for its use in other types of retail businesses. Basic principles of franchising were introduced into retail marketing with events such as the development of the Ben Franklin general merchandise store in 1920 and the emergence of national A & W walk-up root beer stands in 1925. Also in 1925, Howard Johnson offered three flavors of ice cream in a drugstore in Massachusetts. This ice cream business was expanded through franchising to a group of restaurants on the East Coast. The first Howard Johnson® Restaurant appeared on a turnpike in 1940, and the first Howard Johnson Motor Lodge opened in the year 1954. Baskin-Robbins® opened its first ice cream stores in 1940.

Franchising in the 1950s

The decade of the 1950s was a boom period for franchising. Major factors for this growth were an expanding postwar economy and rapid development of the interstate highway system, which encouraged the growth of restaurants, gasoline stations, and other franchises. Many well-known restaurant franchises started during this time. Colonel Harlan Sanders initiated his first Kentucky Fried Chicken franchise in 1950 and built a chain of more than 600 restaurants during the decade. The greatest success story in restaurant franchising is often credited to the salesman Ray Kroc, who sold Multimixers to a small hamburger stand in San Bernardino, California. The owners of this restaurant were Maurice and Dick McDonald. Ray Kroc was impressed by the volume of business at this walk-up stand and encouraged the McDonald brothers to expand their business. He volunteered to franchise the McDonald's® concept and founded the McDonald's Corporation in 1955. Even though this was not the first restaurant franchise in the United States, it was a significant landmark in the history of restaurant franchising. In 1959, the International House of Pancakes® initiated the breakfast concept. Other franchisors, including Dairy Queen®, Orange Julius®, Tastee-Freeze®, and Dunkin' Donuts®, established franchises, mainly along the growing interstate highway network. During 1953–1954, franchising was also introduced in the real estate and hotel industries and began spreading to diverse service industries such as dry cleaning, employment services, and tax accounting. The extent of growth can be summarized by the fact that in 1950 fewer than 100 companies used franchising, whereas by the end of the decade more than 900 companies had franchise operations with an estimated 200,000 franchised outlets.

Franchising in the 1960s

The success and rapid development of franchise chains, coupled with an expanding economy, led to a surge of franchising activity in 1960s. Several companies turned to franchising in order to build national marketing networks. An estimated 100,000 new franchise businesses were initiated between the years 1964 and 1969. An additional 50,000 franchises were initiated between 1969 and 1973. In 1968, the franchise industry recorded sales of over $100 billion, which amounted to more than 10 percent of the U.S. gross national product.

Franchising in the 1970s

The glamour of franchising and the rush to get into this type of business led to numerous franchise offerings that were hastily structured, ill-conceived, poorly capitalized, or, in some cases, blatantly fraudulent. A growing number of public complaints, class action suits, and business failures resulted. Based on the hearings by the Small Business Committees of the U.S. Senate and the House of Representatives, and with the support of the franchising industry, several states adopted disclosure/registration requirements for franchised businesses in early 1970. In 1979, the Federal Trade Commission passed the Franchise Disclosure Act, which identifies the twenty sections that a franchise disclosure document must have. This document, commonly referred to as a *prospectus,* is given to prospective franchisees by the franchisor.

During the 1970s, franchising continued to grow, with business-format franchising adding more than 19,000 franchises between 1976 and 1980. Although the number of units developed in the later years was relatively smaller, there was a significant increase in franchise sales, with retail sales increasing by about 140 percent.

Franchising in the 1980s

A continued emphasis on and growth in franchising emerged as a significant force in the U.S. economy in the 1980s. A steady and impressive growth throughout this decade in both franchise sales and in the number of new franchise businesses was recorded. Despite the recessionary conditions of 1982–1983, the number of business-format franchises increased by approximately 51,000 between 1980 and 1986. By the end of the 1980s, franchising was firmly established as an integral part of U.S. and international business. The close association of franchising with the service sector of the economy is attributed to the insulation of franchise businesses from recessionary conditions in the early 1980s.

Franchising in the 1990s

A variety of economic, demographic, and social factors continues to influence the growth of both the service sector and franchising. The aging of the baby boom generation, the increase in numbers of women entering the workforce, the growing population of active retirees, and the continued rising trend in two-income families are creating a demand for services most logically supplied by franchising. Changing attitudes toward convenience, technological advances, mass advertising methods, emergence of electronic devices for home and business, and an emphasis on quality are all encouraging the development of franchised businesses. Even existing businesses started seeking cost-effective ways of expanding distribution channels in light of the unfavorable economic conditions. A report by the Small Business Committee of the U.S. House of Representative (1990) states that "franchising has provided the means for merging the seemingly conflicting interests of existing businesses with those of aspiring entrepreneurs in a single process that promotes business expansion, entrepreneurial opportunity and shared cost and risk."

A major development in franchising took place in the area of international franchising. Many U.S. franchises began expanding throughout the world. From Europe to South Asia to the Pacific Rim, U.S. franchise companies have found receptive market niches for their products and services. The changing political situation in Eastern Europe, Russia, and other former Soviet republics, the European Common Market development, and the changing economic situation of the Pacific Rim countries, all had a tremendous impact on the U.S. franchise businesses overseas, including restaurant franchises, which were among the first to penetrate in those areas.

Restaurant Study 1

Wendy's International, Inc.

Wendy's began as the dream of Dave Thomas, founder, senior chairman, and "Wendy's Dad." He wanted to build a better hamburger.

His dream became a reality in Columbus, Ohio, in 1969, when Thomas opened the first Wendy's Old Fashioned Hamburgers restaurant, named after one of his daughters, Melinda Lou, nicknamed "Wendy" by her brother and sisters.

Today, there are nearly 5,400 Wendy's restaurants in the United States and 34 countries and territories. Systemwide sales have grown from less than $300,000 in 1970 to $6 billion today. More than five million customers are served each day in Wendy's restaurants.

In December 1995, Wendy's merged with Tim Hortons, the second largest restaurant chain in Canada. Tim Hortons operates as a subsidiary of Wendy's, and continues to independently operate their coffee and fresh baked goods restaurants.

In addition, Wendy's and Tim Hortons developed an innovative "combo" restaurant unit that provides a unique approach to food-service convenience and consumer appeal. The combo buildings are freestanding with a shared dining room and separate food preparation, storage areas, and staffs.

The merger with Tim Hortons expands Wendy's market presence in Canada. Tim Hortons will continue to add new units through Canada, and development plans for combo units call for aggressive expansion in Canada, as well as in the United States. Wendy's have international units in several countries including Japan, Taiwan, and England.

More than 150,000 Wendy's employees worldwide live by the creed that "Quality Is Our Recipe." Wendy's goal is to be better than the rest, with high-quality, great-tasting food, clean, comfortable surroundings and fast, friendly service. Delivering Total Quality is the focus of every aspect of Wendy's operations and the philosophy of every employee. It is this dedication to quality that makes Wendy's a leader in the food service industry and a favorite of American consumers.

Wendy's menu offers customers a variety of great tasting and nutritious choices, and of course, hamburgers remain one of Wendy's most popular items. Wendy's hamburgers are 100 percent pure domestic

(a)

(b)

An exterior of a typical Wendy's restaurant (courtesy Wendy's International, Inc.)

ground beef, cooked to order and served hot-off-the-grill with the customer's choice of toppings. Wendy's Chili—a closely guarded secret recipe—is made from scratch every day in each Wendy's restaurant.

Wendy's was first in the quick-service restaurant industry to utilize the Pick-Up Window, and first to offer baked potatoes and the salad bar nationwide.

As lifestyles have changed over the years, Wendy's has adjusted its menu to reflect our customers changing tastes. In addition to the salad bar, Wendy's offers both a Chicken Fillet Sandwich and a Grilled Chicken sandwich, as well as Fresh Salads to Go: Grilled Chicken, Deluxe-Garden, Taco, Caesar, and Side Salads with a variety of fat-free and reduced fat, reduced calorie salad dressings. New to the Wendy's menu are Fresh Stuffed Pitas, freshly prepared and tasty salads wrapped in a pita bread. The innovative pitas are available in four varieties: Chicken Caesar, Garden Ranch Chicken, Garden Veggie, and Classic Greek.

In 1990, Dave Thomas, who was adopted as a child, became a spokesperson for the White House initiative on adoption, called "Adoption Works . . . For Everyone." In this capacity, Thomas works tirelessly to raise adoption awareness nationwide.

HISTORICAL HIGHLIGHTS

November 15, 1969	First Wendy's Old Fashioned Hamburgers restaurant opened in downtown Columbus, Ohio, at 257 E. Broad St.
November 21, 1970	Second Wendy's opened in Columbus, featuring a Pick-Up Window with a separate grill, a unique feature in the fast-service restaurant industry.
December 1, 1972	Wendy's Management Institute formed to develop management skills in restaurant managers, supervisors, and franchise owners.
December 31, 1974	After 5 years, Wendy's net income exceeded $1 million and revenues topped $13 million. Total restaurant sales were nearly $25 million.
June 25, 1975	100th Wendy's restaurant opened in Louisville, Kentucky.
September 8, 1975	Wendy's stock was first traded on the "over the counter" market (WNDY)
September 23, 1975	First international restaurant opened in Canada.
September, 1976	Wendy's had its first public offering in the amount of one million common shares at $28 per share.
December 15, 1976	Wendy's 500th restaurant opened in Toronto, Canada.
April 1, 1977	Wendy's went national with its first television commercial, making it the first chain with fewer than 1,000 restaurants to launch a national advertising campaign.

March 21, 1978	1,000th Wendy's opened in Springfield, Tennessee. Wendy's has opened 1,000 restaurants in 100 months—an industry record.
March 20, 1979	1,500th Wendy's opened in San Juan, Puerto Rico.
November 15, 1979	Wendy's celebrated its 10th birthday with 1,767 restaurants. Wendy's opened more than 750 restaurants in 21 months, or almost 1.5 per day.
November 1979	The Salad bar was approved as an addition to Wendy's menu. It was the first new product added and began Wendy's menu diversification program.
November 15, 1980	Wendy's opened its 2,000th restaurant.
May 27, 1981	Wendy's was listed on the New York Stock Exchange, using the symbol "WEN."
May 26, 1983	Wendy's 2,500th restaurant opened in Silver Spring, Maryland.
October 15, 1983	Wendy's became the first national chain to introduce Hot Stuffed Baked Potatoes nationwide.
January 9, 1984	Wendy's introduced "Where's the Beef?" starring Clara Peller and sidekicks Elizabeth Shaw and Mildred Lane. "Where's the Beef?" was the most popular commercial of the year.
November 15, 1984	Wendy's celebrated its 15th birthday with more than 2,900 restaurants in the United States and 14 countries.
February 6, 1985	Wendy's opened its 3,000th restaurant in the French Quarter in New Orleans.
May 3, 1985	Wendy's opened a restaurant at the Columbus Zoo, the first quick-service restaurant in a U.S. zoo.
September 1986	Wendy's introduced the Big Classic, a quarter-pound hamburger loaded with fresh toppings, served on a corn-dusted kaiser-style bun. It becomes the Big Bacon Classic.
April 1989	Wendy's introduced the national Dave Thomas Advertising Campaign.
October 1989	Introduction of Wendy's Super Value Menu; nine quality items available every day for only 99 cents each.
Summer 1990	Wendy's announced an expanded corporate nutrition policy with the introduction of a Grilled Chicken Sandwich.

October 1990	Dave Thomas was asked by President George Bush to serve as spokesperson for the national initiative, "Adoption Works . . . For Everyone." Wendy's embraced adoption as its national charitable cause, committing time and financial resources to raise adoption awareness.
September 1991	*Dave's Way,* Dave Thomas's autobiography was published, kicking off a national tour to promote the book. All of Thomas's profits from book sales benefit national adoption awareness programs.
March 1992	Wendy's added five Fresh Salads To Go to the menu: Grilled Chicken, Deluxe Garden, Taco, Caesar, and Side Salads.
September 1992	The first Wendy's Three-Tour Challenge Golf tournament was played to benefit the Dave Thomas Foundation for Adoption.
December 1992	Wendy's opened its 4,000th restaurant in Bentonville, Arkansas.
March 25, 1993	Dave Thomas, 45 years after he dropped out, returned to high school to receive his high school diploma and GED from Coconut Creek High School in Ft. Lauderdale.
September 1994	Dave Thomas published a second book, *Well Done!,* about successful people from all walks of life. All of Thomas's proceeds go to the Dave Thomas Foundation for Adoption.
November 15, 1994	Wendy's celebrates its 25th birthday with nearly 4,400 restaurants in 34 countries and territories. The company finished the year with all-time record sales and profits.
December 1995	Wendy's merged with Tim Hortons, a Canadian restaurant chain. Tim Hortons became a subsdiary of Wendy's and continues to operate their coffee and fresh baked goods restaurants throughout Canada.
September 1996	Dave Thomas filmed his 500th commercial, making this the longest running advertising campaign featuring a company founder as spokesperson.
March 1997	Wendy's opened its 5,000th restaurant, a Wendy's/ Tim Hortons combo unit located in Columbus, Wendy's hometown.

April 1997   Wendy's introduced Fresh Stuffed Pitas, freshly prepared and tasty salads wrapped in a pita bread which can be conveniently eaten like a sandwich. The product comes in four varieties: Chicken Caesar, Garden Ranch Chicken, Garden Veggie, and Classic Greek.

CHAPTER 2

Restaurant Franchising in the U.S. Economy

FRANCHISING IN THE ECONOMY

Franchising continues to be a strong and significant contributor to the American economy. According to the International Franchise Association, sales of goods and services through franchising reached almost $800 billion in the 1990s. Much of this rapid growth is led by business-format franchising. The total employment that can be attributed to franchising, including part-time workers, amounts to over $7.7 million. The volume of sales and the number of units owned or franchised by business-format franchisors have risen steadily since 1972.

Franchise restaurants fall under the category of business-format franchising because product, service, and trademark—the entire concept—are included in franchising. Restaurant franchising continues to offer opportunities for those seeking to own a business, both in the United States and abroad.

Large franchisors (those with 1,000 or more units each) were reported by the International Franchise Association as continuing to dominate business-format franchising, with sixty-nine companies accounting for 51.7 percent of all sales and 53.2 percent of all establishments in 1988. Of these large franchises, sixteen companies, or 23 percent, were engaged in restaurant franchising. Sales of franchised restaurants of all types were expected to reach almost $69 billion in 1989, which is an 8.7 percent increase over 1988 sales of $64 billion. These sales were expected to be $76.5 billion (11.3 percent) in 1990. Meanwhile, the number of franchised restaurants, which totaled 90,345 in 1988, increased to 94,285 in 1989 and was expected to reach 102,135 in 1990. Employment in the franchise restaurant sector was 2,726,605 in 1988, up from 2,453,261 in 1986. Although these figures are rather outdated, the intent of including these data is to illustrate the size of restaurant franchising and its impact on the economy.

The highest numbers of franchised restaurants are in California, Texas, Ohio, Florida, and Illinois. These states have always had a high concentration of franchised restaurants. Among the restaurant menu themes, hamburger seems to be dominant, accounting for approximately half of the total sales and for about 40 percent of the total number of establishments. Franchised pizza restaurants are second in line, with about 20 percent of the total number of establishment. Other popular franchise restaurant menu themes include steak and chicken.

Some of the current trends show that restaurant franchises have adapted to consumer demands by expanding menus, using fresh ingredients, adding salads and salad bars, and offering more takeout and delivery services. According to the 1990 report by the International Franchise Association, restaurant franchisors were trying to strengthen their presence in current market segments rather than expanding into new markets. Consequently, some companies were searching for profitable mergers and acquisitions or even selling off unprofitable units and restructuring corporate operations. International operations of franchised restaurants continue to expand as the domestic market becomes more competitive.

The United States is the leader in restaurant franchising, which contributes significantly to its economy. Restaurant franchising is responsible for creating job opportunities, promoting entrepreneurship, introducing new business services, and providing increased business and export opportunities.

TYPES OF FRANCHISE RESTAURANTS

The International Franchise Association classifies restaurants based on menu theme into the following segments: (1) chicken, (2) hamburger, (3) pizza, (4) Mexican, (5) seafood, (6) pancakes/waffles, (7) steak (full menu), (8) sandwich, and (9) other. *Nation's Restaurant News,* which ranks chains based on a variety of criteria, categorizes them into the following market segments: (1) sandwich, (2) contract, (3) pizza, (4) family, (5) dinner house, (6) hotel, (7) chicken, (8) snack, (9) cafeteria, (10) grill buffet, and (11) others. However, this classification includes all types of chains, including franchised and nonfranchised units. Other classifications are similar to these categories. The major segments and their current significance in the U.S. economy are discussed in this chapter. Figure 2.1 shows the market share of the top one hundred chains, according to *Nation's Restaurant News* research. The sandwich segment occupies 40.9 percent of the market share and thus ranks as the largest segment among restaurants, followed by pizza, family, and dinner house. The chicken segment, which is gaining popularity, occupies 5.5 percent of the market share.

The top one hundred chains, ranked by market share, are shown in Table 2.1. The leaders are McDonald's, Burger King®, Pizza Hut, Taco Bell, Wendy's®, and Kentucky Fried Chicken. The next five are Aramark, Marriott Management Services, SUBWAY, Hardee's®, and Domino's Pizza®.

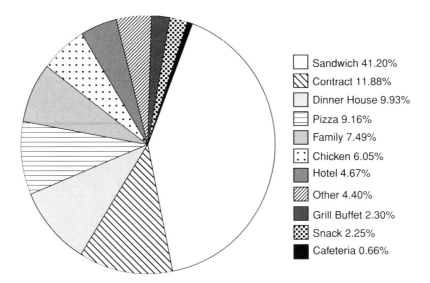

FIGURE 2.1 Top 100 market share by segments (1998) (Adapted from Nations' Restaurant News Special Report).

GROWTH IN FRANCHISE RESTAURANTS

The top one hundred chains, ranked by number of units by the *Nation's Restaurant News,* is shown in Table 2.2. It should be noted that while some of these have all company-owned units with no franchise units, others have some units that are company owned and some that are franchised. McDonald's has 12,380 units, out of which 10,582, or 85 percent, are franchised units. Pizza Hut has a total of 8,698 units, out of which 4,875 (56 percent) are franchised units. The top one hundred chains ranked by growth in U.S. franchise units are presented in Table 2.3. The top five are Bennigans, Ruby Tuesday, Coco's, Starbucks, and Churchs Chicken. A noteworthy feature is that some of the leading restaurants, such as McDonald's, were ranked 39TH in the number of growth in units. This is indicative of a change in strategy of larger chains, a lack of prime locations in the United States, the growth of relatively newer restaurants with different concepts, or a combination of these.

TABLE 2.1 TOP 100 CHAINS RANKED BY U.S. SYSTEMWIDE FOODSERVICE SALES

Latest Year Rank	Preceding Year Rank	Chain	Concept	Parent Company	Fiscal Year End	U.S. Systemwide Foodservice Sales (by Fiscal Year, in Millions)		
						Latest**	Preceding	Prior
1	1	McDonald's	Sandwich	McDonald's Corp.*	Dec. '97	$17,124.7	$16,369.6	$15,905.0
2	2	Burger King	Sandwich	Diageo PLC	Sept. '97	7,900.0	7,300.0	6,700.0
3	3	Pizza Hut	Pizza	Tricon Global Restaurants Inc.*	Dec. '97	4,700.0	4,927.0	5,300.0
4	4	Taco Bell	Sandwich	Tricon Global Restaurants Inc.*	Dec. '97	4,650.0	4,575.0	4,600.0
5	5	Wendy's	Sandwich	Wendy's International Inc.*	Dec. '97	4,560.0	4,360.3	4,138.2
6	6	KFC	Chicken	Tricon Global Restaurants Inc.*	Dec. '97	4,000.0	3,900.0	3,700.0
7	8	Aramark Global Food/Leisure Services	Contract	Aramark Corp.	Sept. '97	3,270.0	3,029.0	2,782.0
8	9	Marriott Management Services	Contract	Marriott International Inc.*	Dec. '97	3,250.0	2,750.0	3,100.0
9	10	Subway	Sandwich	Doctor's Associates Inc.	Dec. '97	2,869.5	2,700.0	2,600.0
10	7	Hardee's	Sandwich	CKE Restaurants Inc.*	Dec. '97	2,650.0	3,055.0	3,325.0
11	11	Domino's Pizza	Pizza	Domino's Pizza Inc.	Dec. '97	2,480.0	2,300.0	2,100.0
12	12	Arby's	Sandwich	Triarc Corp.*	Dec. '97	2,000.0	1,867.0	1,817.0
13	13	Denny's	Family	Advantica Restaurant Group Inc.*	Dec. '97	1,872.0	1,850.0	1,785.0

TABLE 2.1 (continued)

Latest Year Rank	Preceding Year Rank	Chain	Concept	Parent Company	Fiscal Year End	U.S. Systemwide Foodservice Sales (by Fiscal Year, in Millions)		
						Latest*	Preceding	Prior
14	17	Applebee's Neighborhood Grill & Bar	Dinner House	Applebee's International Inc.*	Dec. '97	1,800.0	1,523.0	1,242.0
14	14	Red Lobster	Dinner House	Darden Restaurants Inc.*	May '98	1,800.0	1,774.9	1,814.0
16	15	Dunkin' Donuts	Snack	Allied Domecq PLC	Aug. '97	1,750.0	1,568.8	1,436.6
17	16	LSG Lufthansa Service/Sky Chef	Contract	Onex Corp.	Dec. '97	1,600.0	1,530.0	739.0
18	18	Little Caesars Pizza	Pizza	Little Caesar Enterprises	Dec. '97	1,375.0	1,400.0	1,450.0
18	20	Olive Garden	Dinner House	Darden Restaurants Inc.*	May '98	1,375.0	1,255.0	1,228.1
20	23	Jack in the Box	Sandwich	Foodmaker Inc.*	Oct. '97	1,352.8	1,200.6	1,102.4
21	19	Marriott hotels, resorts & suites	Hotel	Marriott International Inc.*	Dec. '97	1,310.0	1,298.0	1,282.0
22	21	Dairy Queen	Sandwich	International Dairy Queen Inc.*	Nov. '97	1,250.0	1,225.0	1,200.0
23	26	Outback Steakhouse	Dinner House	Outback Steakhouse Inc.*	Dec. '97	1,246.0	1,017.0	798.4
24	24	Boston Market	Chicken	Boston Chicken Inc.*	Dec. '97	1,197.0	1,099.0	755.0
25	25	Chili's Grill & Bar[5]	Dinner House	Brinker International Inc.*	June '98	1,195.6	1,035.1	936.3

26	21	Shoney's	Family	Shoney's Inc.*	Oct. '97	1,150.0	1,225.0	1,250.0
27	27	Sonic Drive-In	Sandwich	Sonic Corp.*	Aug. '97	1,142.6	984.8	880.5
28	28	T.G.I Friday's	Dinner House	Carlson Cos. Inc.	Dec. '97	986.0	935.4	872.2
29	29	7-Eleven	C-Store	Southland Corp.*	Dec. '97	958.0	907.2	850.0
30	30	Hilton Hotels	Hotel	Hilton Hotels Corp.*	Dec. '97	905.0	892.0	882.0
31	32	Canteen Services[4]	Contract	Compass Group PLC	Dec. '97	900.0	850.0	805.0
32	31	Sheraton Hotels	Hotel	ITT Corp.*	Dec. '97	890.0	852.5	850.0
33	46	Papa John's Pizza	Pizza	Papa John's International Inc.*	Dec. '97	867.6	619.2	458.7
34	37	Cracker Barrel Old Country Store	Family	Cracker Barrel Old Country Store Inc.*	July '97	863.1	733.9	610.0
35	35	Intl. House of Pancakes/IHOP Restaurants	Family	IHOP Corp.*	Dec. '97	850.0	750.0	675.0
36	33	Long John Silver's	Fish QSR	Long John Silver's Restaurants	June '98	843.0	848.0	902.0
37	43	Eurest Dining Services[4]	Contract	Compass Group PLC	Dec. '97	825.0	660.0	625.0
38	34	Sodexho U.S.A.	Contract	Sodexho SA	Aug. '97	820.0	780.0	730.0
39	35	Dobbs International Services	Contract	Viad Corp.*	Dec. '97	800.0	750.0	735.0
40	39	Golden Corral	Grill-Buffet	Investors Management Corp.	Dec. '97	771.0	710.9	618.1
41	38	Big Boy Restaurant & Bakery	Family	Elias Bros. Restaurants Inc.	Dec. '97	750.0	725.0	865.2
42	42	Popeyes	Chicken	AFC Enterprises	Dec. '97	719.7	666.0	656.0

27

Table 2.1 (continued)

Latest Year Rank	Preceding Year Rank	Chain	Concept	Parent Company	Fiscal Year End	U.S. Systemwide Foodservice Sales (by Fiscal Year, in Millions)		
						Latest	Preceding	Prior
43	41	Perkins Family Restaurants	Family	The Restaurant Co.	Dec. '97	711.0	678.0	633.9
44	45	Carl's Jr.	Sandwich	CKE Restaurants Inc.*	Jan. '98	685.0	625.0	560.0
45	44	Holiday Inns	Hotel	Bass PLC	Oct. '97	658.0	650.0	630.0
46	40	Ponderosa Steakhouse	Grill-Buffet	Metromedia Co.	Dec. '97	657.0	680.3	748.8
47	52	Chick-fil-A	Chicken	Chick-fil-A Inc.	Dec. '97	643.2	570.0	501.6
48	54	Ruby Tuesday	Dinner House	Ruby Tuesday Inc.*	May '98	640.0	556.9	516.7
49	48	Ryan's Family Steak House	Grill-Buffet	Ryan's Family Steak Houses Inc.*	Dec. '97	636.0	604.0	560.0
50	47	Disney theme parks, hotels & resorts	Theme Park	Walt Disney Co.*	Sept. '97	627.0	615.0	601.0
51	51	Bob Evans Restaurants	Family	Bob Evans Farms Inc.*	April '98	618.0	575.0	550.0
52	49	Friendly's Ice Cream	Family	Friendly Ice Cream Corp.*	Dec. '97	608.2	598.7	593.6
53	50	Baskin-Robbins	Snack	Allied Domecq PLC	Aug. '97	605.0	576.7	649.0
54	65	Starbucks	Coffee	Starbucks Corp.*	Sept. '97	595.0	435.0	305.0
55	55	Churchs Chicken	Chicken	AFC Enterprises	Dec. '97	573.9	829.0	601.0
56	53	Old Country Buffet	Buffet	Buffets Inc.*	Dec. '97	550.0	560.0	623.3

57	57	Radisson	Hotel	Carlson Cos. Inc.	Dec. '97	525.0	490.0	455.0
58	57	Waffle House	Family	Waffle House Inc.	Dec. '97	625.0	525.0	512.2
59	59	Lone Star Steakhouse & Saloon	Dinner House	Lone Star Steakhouse & Saloon Inc.*	Dec. '97	509.6	446.3	327.0
60	58	Ramada Inn	Hotel	Hospitality Franchise Systems	Dec. '97	500.0	489.0	481.0
61	61	Luby's Cafeteria	Cafeteria	Luby's Cafeterias Inc.*	Aug. '97	495.4	450.1	419.0
62	60	Bennigan's	Dinner House	Metromedia Co.	Dec. '97	487.0	458.0	453.0
63	59	Captain D's Seafood	Fish QSR	Shoney's Inc.*	Oct. '97	464.4	480.0	475.0
64	64	Hyatt Hotels	Hotel	Hyatt Hotels Corp.	Dec. '97	460.0	440.0	415.0
65	63	Whataburger	Sandwich	Whataburger Inc.	Sept. '97	459.0	445.0	674.0
66	67	The Wood Co.	Contract	The Albert Abela Group	Dec. '97	440.0	400.0	330.0
67	68	Ogden Entertainment Services	Contract	Ogden Corp.*	Dec. '97	425.9	391.9	301.3
68	115	Charwells[4]	Contract	Compass Group PLC	Dec. '97	425.0	210.0	120.0
69	66	Sbarro, The Italian Eatery	Pizza	Sbarro Inc.*	Dec. '97	421.7	400.2	394.8
70	70	Round Table Pizza	Pizza	Round Table Franchise Corp.	June '98	391.3	384.8	373.8
71	71	White Castle	Sandwich	White Castle System Inc.	Dec. '97	384.7	351.1	325.4
72	73	ServiceMaster Food Mgmt. Services	Contract	Service Master L.P.*	Dec '97	350.0	342.6	332.0
73	82	Blimpie Subs & Salads	Sandwich	Blimpie International*/ Blimpie Associates	June '98	345.0	287.0	231.0
74	69	Sizzler	Grill-Buffet	Sizzler International Inc.*	April '98	340.0	388.0	580.0

TABLE 2.1 (continued)

Latest Year Rank	Preceding Year Rank	Chain	Concept	Parent Company	Fiscal Year End	U.S. Systemwide Foodservice Sales (by Fiscal Year, in Millions)		
						Latest	Preceding	Prior
75	75	Hooters	Dinner House	Hooters of America Inc.	Dec. '97	339.0	327.0	301.0
76	81	Krystal	Sandwich	Krystal Co.*	Dec. '97	330.0	292.0	288.0
77	86	Steak 'n Shake	Family	Consolidated Products Inc.*	Sept. '97	323.7	275.0	215.0
78	72	Host International (proprietary)	Contract (airport)	Host Marriott Services Corp.*	Dec. '97	317.5	345.7	329.5
79	74	Checkers Drive-In	Sandwich	Checkers Drive-In Restaurants Inc.*	Dec. '97	310.3	328.0	388.9
80	76	Western Sizzlin'	Grill Buffet	Western Sizzlin' Corp.	Dec '97	302.0	322.0	335.0
81	86	KCafe/Eatery Express (Kmart)	In-Store	Kmart Corp.*	Jan. '98	300.0	275.0	265.0
81	78	Restaura Inc.	Contract	Viad Corp.*	Dec. '97	300.0	310.0	310.0
81	78	TCBY	Snack	TCBY Enterprises Inc.*	Nov. '97	300.0	310.0	315.0
84	95	HomeTown Buffet	Buffet	Buffets Inc.*	Dec. '97	295.0	250.0	206.0
85	89	Chuck E. Cheese's	Pizza	ShowBiz Pizza Time Inc.*	Dec. '97	292.3	268.0	252.0
86	77	Rally's Hamburgers	Sandwich	Rally's Hamburgers Inc.*	Dec '97	290.1	316.7	360.0
87	104	Romano's Macaroni Grill[5]	Dinner House	Brinker International Inc.*	June '98	289.9	221.5	176.7

88	84	Coco's	Family	Advantica Restaurant Group Inc.*	Dec. '97	288.0	278.5	290.2
89	88	Piccadilly Cafeteria	Cafeteria	Piccadilly Cafeterias Inc.*	June '98	280.0	273.8	272.4
89	100	Wal-Mart snack bars, including Radio Grill	In-Store	Wal-Mart Stores Inc.*	Jan. '98	280.0	235.0	212.0
91	107	Fazoli's	Italian QSR	Seed Restaurant Group Inc.	March '98	275.5	220.7	137.0
92	150	Fine Host	Contract	Fine Host Corp.*	Dec. '97	275.1	144.4	107.9
93	94	Marie Callender's Pie Shops	Family	Wilshire Restaurant Group Inc.	Dec. '97	272.0	254.0	250.0
94	90	Godfather's Pizza	Pizza	Godfather's Pizza Inc.	May '98	270.8	265.5	260.0
95	97	Pizzeria Uno	Dinner House	Uno Restaurant Corp.*	Sept. '97	269.6	245.9	228.3
96	98	Tony Roma's Famous for Ribs	Dinner House	NPC International Inc.*	March '98	265.0	239.4	229.3
97	93	Stuart Anderson's Black Angus	Dinner House	American Restaurant Group Inc.	Dec. '97	264.2	254.9	243.6
98	118	Schlotzsky's Deli	Sandwich	Schlotzsky's Inc.*	Dec. '97	263.8	197.6	140.0
99	80	Red Robin Burger & Spirits Emporium	Dinner House	Red Robin International Inc.	Dec. '97	261.8	293.8	257.0
100	92	Westin Hotels & Resorts	Hotel	Starwood Capital Group LP and partners	Dec. '97	260.0	259.0	254.0
TOTALS:						$117,870.5	$111,438.7	$106,517.7

TABLE 2.1 (continued)

* Publicly held U.S. company
** Actual result, estimate or projection
Source: NRN Research
Explanation of Column Headings

Latest-Year Rank: The Top 100 chains and separately listed Top 100 Companies are ranked in descending order of magnitude on the basis of actual, estimated or projected results for the organizations latest fiscal years, generally the year ended or ending nearest Dec. 31, 1997. Data are limited to the chains' systemwide food and beverage sales in the United States only; company data are limited to domestically generated revenue derived from foodservice only. Chains and companies with tied results in any Top 100 ranking are assigned the same numeric rank.

Preceding-Year Rank: The same 100 chains and separately listed Top 100 Companies are ranked on the basis of U.S. sales and revenue in their preceding full fiscal year. Chains and companies assigned a preceding-year rank beyond No. 100 would have held that position in the preceding year among the "Second 100" chains and companies, whose results will appear in a companion study in the July 20 issue of Nation's Restaurant News. Chains and companies with tied results in any Top 100 ranking are assigned the same numeric rank.

Fiscal year-end: The ending date of the year represented in the "Latest" year column. If an entity's actual ending date occurs in the first half of a month, the preceding full month is shown as its fiscal year-end, pursuant to conventional financial-reporting practices.

"Latest" "Preceding" and "Price": Three consecutive years, the latest of which ended or ends in the month and year indicated in the corresponding "Fiscal Year End" column.

"Top 100 Chains" Column Headings Definitions:

Chain: The brand name of the restaurants, hotels, contract foodservice systems, retail stores or other entities in a multiunit organization, as identified by its signs, logotypes and trademarks.

Concept: The type of restaurant or foodservice operation run by the chain, as defined generically by its food type, service style, retail context or operating format. Most chain concepts are self-defining and are categorized on the basis of the Top 100 study's traditional categorical parameters. Full-service concepts other than hotel operations, that is, dinner houses and family restaurants, are defined as such and distinguished from one another on the basis of their generally dissimilar price-point ranges and by the fact that most dinner houses have full bar operations but do not serve breakfast, whereas most family restaurants do not have bars but do serve breakfast.

Parent Company: The business entity that owns a chain's trademarks and master franchising rights, whether directly or through an operating subsidiary, and which ultimately profits from operating and/or franchising or licensing its concept. An asterisk following a company's name indicates that it is a U.S.-based entity whose stock is publicly traded; foreign-based companies not so indicated may be public in their home countries.

U.S. Systemwide Foodservice Sales: Foodservice sales at all restaurants, hotels, stores, contract locations of other outlets in a chain, including company-owned, company-managed, franchised and licensed units. Sales from manufacturing, distribution, merchandise, room revenues, video games and other nonfood sources of income are excluded.

"Top 100 Companies" Column Heading Definitions:

Company: The business entity that ultimately owns the specified foodservice revenue; not an owned subsidiary that may function as "parent company" to a chain or foodservice division. An asterisk following a company's name indicates that it is a U.S.-based entity whose stock is publicly traded; foreign-based companies not so indicated may be public in their home countries.

Chains, divisions: The names of the chains or foodservice groups operated or franchised by the company or its operating subsidiary. In some instances the listed chain or chains are contract foodservice systems, hotel chains, theme-park groups or franchises of restaurant brands owned by other companies.

U.S. Food & Beverage Revenue: Domestically generated foodservice dollars collected by the company. Included are sales at company-owned and profit-and-loss-contracted operations as well as food-and-beverage sales royalties and fees collected from franchises. Excluded from revenue figures are total sales at franchised restaurants, revenues generated by manufacturing or wholesaling of food or other products; foodservice distribution revenues; and other revenues from nonfoodservice sales at restaurants, such as merchandise and video games. Figures for hotels, contractors, in-store feeders and theme parks should exclude nonfoodservice retail sales and nonfoodservice contract activities, such as income from rooms, souvenirs, property maintenance, ticket sales and grocery or other general retail.

TABLE 2.2 TOP 100 CHAINS RANKED BY LATEST-YEAR TOTAL NUMBER OF U.S. UNITS

					Number of Units							
				Latest			Preceding			Prior		
Latest Year Rank	Preceding Year Rank	Chain	Fiscal Year End	Total	Company	Franchised	Total	Company	Franchised	Total	Company	Franchised
1	1	Canteen Services	Dec. '97	17,000	17,000	0	16,050	16,050	0	15,200	15,200	0
2	2	McDonald's	Dec. '97	12,380	1,798	10,582	12,094	1,847	10,247	11,368	1,839	9,529
3	3	Subway	Dec. '97	11,165	0	11,155	10,848	0	10,848	10,093	0	10,093
4	4	Pizza Hut	Dec. '97	8,698	3,823	4,875	8,910	4,818	4,092	8,883	5,201	3,682
5	5	Burger King	Sept. '97	7,414	512	6,902	6,925	516	6,409	6,492	448	6,044
6	6	Taco Bell	Dec. '97	6,768	2,149	4,619	6,670	2,696	3,974	6,490	3,133	3,357
7	8	7-Eleven	Dec. '97	5,270	2,077	3,193	5,044	2,079	2,965	4,973	2,077	2,896
8	7	KFC	Dec. '97	5,120	1,850	3,270	5,079	1,957	3,122	5,152	2,031	3,121
9	9	Dairy Queen	Nov. '97	5,052	34	5,018	5,035	34	5,001	5,000	0	5,000
10	10	Wendy's	Dec. '97	4,575	1,073	3,502	4,369	1,191	3,178	4,197	1,200	2,997
11	11	Domino's Pizza	Dec. '97	4,431	760	3,671	4,300	720	3,580	4,242	698	3,544
12	12	Little Caesars Pizza	Dec. '97	3,900	670	3,230	4,008	674	3,334	4,187	690	3,497
13	14	Dunkin' Donuts	Aug. '97	3,342	12	3,330	3,188	12	3,176	3,043	18	3,025
14	13	Hardee's	Dec. '97	2,944	863	2,081	3,225	808	2,417	3,395	864	2,531
15	16	Arby's	Dec. '97	2,913	0	2,913	2,859	339	2,520	2,798	366	2,432
16	15	Marriott Management Services	Dec. '97	2,845	2,845	0	2,878	2,878	0	3,100	3,100	0
17	17	Baskin-Robbins	Aug. '97	2,655	5	2,650	2,562	5	2,557	2,504	5	2,499
18	19	TCBY	Nov. '97	2,579	2	2,577	2,497	2	2,495	2,533	42	2,491
19	18	Aramark Global Food/Leisure Services	Sept. '97	2,520	2,520	0	2,550	2,550	0	2,400	2,400	0
20	21	Blimpie Subs & Salads	June '98	2,175	0	2,175	1,800	0	1,800	1,581	0	1,581
21	20	KCafe/Eatery Express (Kmart snack bars)	Jan. '98	2,136	2,136	0	2,134	2,134	0	2,161	2,161	0

#	#	Name	Date									
22	24	Eurest Dining Services	Dec. '97	2,000	2,000	0	1,600	1,600	0	1,515	1,515	0
23	24	Wal-Mart snack bars (incl. Radio Grill)	Jan. '98	1,882	1,882	0	1,724	1,724	0	1,577	1,577	0
24	22	Sodexho U.S.A.	Aug. '97	1,800	1,800	0	1,580	1,580	0	1,519	1,519	0
25	23	Holiday Inns	Oct. '97	1,759	79	1,680	1,719	77	1,642	1,697	70	1,627
26	26	Sonic Drive-In	Aug. '97	1,680	256	1,424	1,567	231	1,336	1,464	178	1,286
27	27	Denny's	Dec. '97	1,592	884	708	1,541	884	657	1,474	923	551
28	30	Papa John's Pizza	Dec. '97	1,517	401	1,116	1,160	303	857	878	217	661
29	29	Jack in the Box	Oct. '97	1,317	963	354	1,244	879	365	1,231	863	368
30	28	Long John Silver's	June '98	1,309	814	495	1,413	928	485	1,476	971	505
31	31	Boston Market	Dec. '97	1,166	307	859	1,087	195	982	829	3	826
32	33	Starbucks	Sept. '97	1,108	932	176	1,009	919	90	751	676	75
33	32	Waffle House	Dec. '97	1,080	530	550	1,080	530	550	1,015	505	510
34	34	Church's Chicken	Dec. '97	1,070	480	590	989	622	367	964	589	375
35	35	Popeyes	Dec. '97	945	119	826	892	120	772	884	117	767
36	37	Applebee's Neighborhood Grill & Bar	Dec. '97	944	190	754	810	148	662	660	128	532
37	38	Sbarro, The Italian Eatery	Dec. '97	805	623	182	757	597	160	725	571	154
38	36	Shoney's	Oct. '97	770	489	281	844	544	300	882	356	526
39	39	Ramada Inn	Dec. '97	750	0	750	730	0	730	700	0	700
40	40	Chick-fil-A	Dec. '97	749	636	113	717	633	84	657	590	67
41	42	Big Boy Restaurant & Bakery	Dec. '97	700	100	600	700	100	600	860	100	760
42	43	Intl. House of Pancakes/IHOP Restaurants	Dec. '97	699	66	633	651	57	594	600	49	551
43	41	Friendly's Ice Cream	Dec. '97	696	662	34	707	707	0	735	735	0
44	45	Carl's Jr.	Jan. '98	686	443	243	643	415	228	633	394	239
45	48	Schlotzsky's Deli	Dec. '97	653	7	646	557	4	553	463	2	461
46	44	Red Lobster	May '98	647	647	0	650	650	0	676	676	0
47	47	Round Table Pizza	June '98	603	23	580	579	15	564	559	15	544

TABLE 2.2 (continued)

Latest Year Rank	Preceding Year Rank	Chain	Fiscal Year End	Latest			Number of Units Preceding			Prior		
				Total	Company	Franchised	Total	Company	Franchised	Total	Company	Franchised
48	46	Captain D's Seafood	Oct. '97	591	378	213	598	379	219	608	310	298
49	70	Chartwells	Dec. '97	580	580	0	285	285	0	160	160	0
50	49	Godfather's Pizza	May '98	554	180	374	540	153	387	525	150	375
51	51	Whataburger	Sept. '97	534	284	250	516	291	225	511	352	159
52	52	Chilli's Grill & Bar	June '98	520	403	117	486	383	103	437	339	98
53	50	Ponderosa Steakhouse	Dec. '97	517	215	302	531	222	309	637	276	361
54	53	Checkers Drive-In	Dec. '97	478	230	248	478	232	246	499	242	257
55	57	Rally's Hamburgers	Dec. '97	477	256	221	444	214	230	481	239	242
56	56	Marriott hotels, resorts & suites	Dec. '97	466	398	68	450	386	64	439	379	60
57	54	Olive Garden	May '98	461	461	0	461	461	0	471	471	0
58	55	Perkins Family Restaurants	Dec. '97	460	136	324	453	134	319	450	139	311
59	60	Outback Steakhouse	Dec. '97	459	373	86	373	318	55	297	258	39
60	58	Golden Corral	Dec. '97	454	185	269	440	196	244	439	223	216
61	59	Bob Evans Restaurants	April '98	393	393	0	381	381	0	363	363	0
62	62	Fine Host	Dec. '97	375	375	0	341	341	0	95	95	0
63	61	The Wood Co.	Dec. '97	370	370	0	352	352	0	311	311	0
64	64	Ruby Tuesday	May '98	357	320	37	328	325	3	301	301	0
65	63	Krystal	Dec. '97	349	248	101	338	249	89	336	256	80
66	65	T.G.I. Friday's	Dec. '97	343	147	196	322	141	181	296	140	156
67	66	Chuck E. Cheese's	Dec. '97	312	246	56	314	244	70	315	226	89
68	67	White Castle	Dec. '97	310	310	0	305	305	0	296	296	0

69	72	Cracker Barrel Old Country Store	July '97	307	307	0	267	267	0	218	218	0
70	74	Fazoli's	March '98	306	187	119	258	166	92	186	124	62
71	68	ServiceMaster Food Management Services	Dec. '97	300	300	0	295	295	0	285	285	0
72	69	Ryan's Family Steak House	Dec. '97	270	245	25	286	261	25	257	231	26
73	81	Lone Star Steakhouse & Saloon	Dec. '97	265	265	0	205	205	0	160	160	0
74	71	Sizzler	April '98	256	66	190	268	69	199	317	87	230
75	75	Western Sizzlin'	Dec. '97	250	16	234	257	7	250	279	6	273
76	80	Steak 'n Shake	Sept. '97	249	194	55	208	161	47	171	137	34
77	73	Old Country Buffet	Dec. '97	242	237	5	261	256	5	249	243	6
78	76	Hilton Hotels	Dec. '97	240	60	180	236	59	177	222	58	164
79	77	Sheraton Hotel	Dec. '97	233	58	175	231	58	173	236	58	178
80	82	Luby's Cafeteria	Aug. '97	229	229	0	204	204	0	187	187	0
81	78	Bennigan's	Dec. '97	228	190	38	221	218	3	221	221	0
82	79	Radisson	Dec. '97	214	14	200	210	12	198	209	7	202
83	84	Hooters	Dec. '97	203	61	142	193	58	135	178	55	123
84	85	Coco's	Dec. '97	195	178	17	188	183	5	194	188	6
85	83	Restaura Inc	Dec. '97	160	160	0	200	200	0	200	200	0
86	86	Pizzeria Unc	Sept. '97	158	92	66	149	86	63	138	79	59
87	87	Marie Callender's Pie Shops	Dec. '97	156	97	59	147	89	58	144	84	60
88	89	Tony Roma's Famous for Ribs	March '98	141	45	96	133	40	93	135	33	102
89	90	Piccadilly Cafeteria	June '98	134	134	0	129	129	0	130	130	0
90	91	Ogden Entertainment Services	Dec. '97	127	127	0	120	120	0	110	110	0
91	92	Home Town Buffet	Dec. '97	116	97	19	109	90	19	89	70	19
92	93	Hyatt Hotels	Dec. '97	112	110	2	105	103	2	100	98	2
93	88	Red Robin Burger & Spirits Emporium	Dec. '97	110	46	64	139	52	87	136	52	84

TABLE 2.2 (continued)

Latest Year Rank	Preceding Year Rank	Chain	Fiscal Year End	Number of Units								
				Latest			Preceding			Prior		
				Total	Company	Franchised	Total	Company	Franchised	Total	Company	Franchised
94	95	Romano's Macaroni Grill	June '98	105	105	0	82	82	0	61	61	0
95	94	Stuart Anderson's Black Angus	Dec. '97	102	102	0	101	101	0	95	95	0
96	96	Dobbs International Services	Dec. '97	100	100	0	80	80	0	70	70	0
97	97	Host International (proprietary)	Dec. '97	63	63	0	65	65	0	68	68	0
98	98	LSG Lufthansa Service/Sky Chef	Dec. '97	56	56	0	56	56	0	34	34	0
99	99	Westin Hotels & Resorts	Dec. '97	48	38	10	44	35	9	42	33	9
100	100	Disney theme parks, hotels & resorts	Sept. '97	21	21	0	21	21	0	21	21	0
		TOTALS:		159,865	65,950	93,915	154,179	65,502	88,677	148,625	63,841	84,784

* Actual results, estimates or projections
Note: For hotels and contractors,
Source: NRN Research

TABLE 2.3 TOP 100 CHAINS RANKED BY GROWTH IN U.S. FRANCHISE UNITS
(YEAR-TO-YEAR PERCENTAGE CHANGE)

				Growth in Franchise Units	
Latest Year Rank	Preceding Year Rank	Chain	Fiscal Year End	Latest* vs. Preceding	Preceding vs. Prior
1	—	Bennigan's	Dec. '97	1,166.67%	—%
2	—	Ruby Tuesday	May '98	1,133.33	—
3	65	Coco's	Dec. '97	240.00	−16.67
4	8	Starbucks	Sept. '97	95.56	20.00
5	51	Churchs Chicken	Dec. '97	60.76	−2.13
6	3	Outback Steakhouse	Dec. '97	56.36	41.03
7	6	Chick-fil-A	Dec. '97	34.52	25.37
8	5	Papa John's Pizza	Dec. '97	30.22	29.65
9	1	Fazoli's	March '98	29.35	48.39
10	14	Blimpie Subs & Salads	June '98	20.83	13.85
11	17	Pizza Hut	Dec. '97	19.13	11.14
12	4	Steak 'n Shake	Sept. '97	17.02	38.24
13	9	Schlotzsky's Deli	Dec. '97	16.82	19.96
14	12	Taco Bell	Dec. '97	16.23	18.38
15	34	Arby's	Dec. '97	15.60	3.62
16	7	Applebee's Neighborhood Grill & Bar	Dec. '97	13.90	24.44
17	31	Sbarro, The Italian Eatery	Dec. '97	13.75	3.90
18	28	Chili's Grill & Bar	June '98	13.59	5.10
19	16	Krystal	Dec. '97	13.48	11.25
20	46	Westin Hotels & Resorts	Dec. '97	11.11	0.00
20	2	Whataburger	Sept. '97	11.11	41.51
22	15	Golden Corral	Dec. '97	10.25	12.96
23	26	Wendy's	Dec. '97	10.20	6.04
24	13	T.G.I. Friday's	Dec. '97	8.29	16.03
25	10	Denny's	Dec. '97	7.76	19.24
26	26	Burger King	Sept. '97	7.69	6.04
26	38	7-Eleven	Dec. '97	7.69	2.38
28	42	Popeyes	Dec. '97	6.99	0.65

TABLE 2.3 (continued)

Latest Year Rank	Preceding Year Rank	Chain	Fiscal Year End	Growth in Franchise Units	
				Latest* vs. Preceding	Preceding vs. Prior
29	32	Sonic Drive-In	Aug. '97	6.59	3.89
30	58	Carl's Jr.	Jan. '98	6.58	−4.60
31	21	International House of Pancakes/IHOP Restaurants	Dec. '97	6.57	7.80
32	25	Marriott hotels, resorts & suites	Dec. '97	6.25	6.67
33	18	Hooters	Dec. '97	5.19	9.76
34	29	Dunkin' Donuts	Aug. '97	4.85	4.99
35	24	Pizzeria Uno	Sept. '97	4.76	6.78
36	44	KFC	Dec. '97	4.74	0.03
37	39	Baskin-Robbins	Aug. '97	3.64	2.32
38	43	TCBY	Nov. '97	3.29	0.16
39	22	McDonald's	Dec. '97	3.27	7.53
40	62	Tony Roma's Famous for Ribs	March '98	3.23	−8.82
41	23	Subway	Dec. '97	2.92	7.48
42	33	Round Table Pizza	June '98	2.84	3.68
43	30	Ramada Inn	Dec. '97	2.74	4.29
44	40	Domino's Pizza	Dec. '97	2.54	1.02
45	41	Holiday Inns	Oct. '97	2.31	0.92
46	55	Long John Silver's	June '98	2.06	−3.96
47	53	Marie Callender's Pie Shops	Dec. '97	1.72	−3.33
48	19	Hilton Hotels	Dec. '97	1.69	7.93
49	37	Perkins Family Restaurants	Dec. '97	1.57	2.57
50	52	Sheraton Hotels	Dec. '97	1.16	−2.81
51	50	Radisson	Dec. '97	1.01	−1.98
52	56	Checkers Drive-In	Dec. '97	0.81	−4.28
53	45	Dairy Queen	Nov. '97	0.34	0.02
54	67	Big Boy Restaurant & Bakery	Dec. '97	0.00	−21.05

TABLE 2.3 (continued)

Latest Year Rank	Preceding Year Rank	Chain	Fiscal Year End	Growth in Franchise Units	
				Latest* vs. Preceding	Preceding vs. Prior
54	46	Home Town Buffet	Dec. '97	0.00	0.00
54	46	Hyatt Hotels	Dec. '97	0.00	0.00
54	65	Old Country Buffet	Dec. '97	0.00	−16.67
54	54	Ryan's Family Steak House	Dec. '97	0.00	−3.85
54	20	Waffle House	Dec. '97	0.00	7.84
60	64	Ponderosa Steakhouse	Dec. '97	−2.27	−14.40
61	69	Captain D's Seafood	Oct. '97	−2.74	−26.51
62	49	Jack in the Box	Oct. '97	−3.01	−0.82
63	59	Little Caesars Pizza	Dec. '97	−3.12	−4.66
64	36	Godfather's Pizza	May '98	−3.36	3.20
65	60	Rally's Hamburgers	Dec. '97	−3.91	−4.96
66	63	Sizzler	April '98	−4.52	−13.48
67	68	Chuck E. Cheese's	Dec. '97	−5.71	−21.35
68	70	Shoney's	Oct. '97	−6.33	−42.97
69	61	Western Sizzlin'	Dec. '97	−6.40	−8.42
70	11	Boston Market	Dec. '97	−12.53	18.89
71	57	Hardee's	Dec. '97	−13.90	−4.50
72	35	Red Robin Burger & Spirits Emporium	Dec. '97	−26.44	3.57

* Actual results, estimates or projections
Note: The 28 Top 100 chains not represented on this table do not franchise; Bennigan's and Ruby Tuesday did not franchise in the "Prior" year.
Source: NRN Research

Franchise restaurants, like other businesses, will increasingly become high-tech to meet the demands of the consumers. This is already evident by the fact that many fast-food restaurants are installing fax machines and stock market price displays in their restaurants in order to meet the demands of the technology-friendly consumer. Phones are being installed on dining tables for easy access to outside businesses. Automatic teller machines are being installed within restaurants. Credit cards are being accepted, and business is being conducted via the Internet.

FRANCHISE RESTAURANT SEGMENTS

It is evident from data obtained by different analysts that the restaurant franchise industry faced some tough challenges in the 1990s. Franchisors had to adapt to these challenges by diversifying, consolidating, introducing new products and services, or other innovative steps. An abbreviated summary of the performance and trends of the different franchise segments follows, with the intent of illustrating the respective roles of these segments in the nation's economy.

Sandwich Restaurants

The growth of sandwich restaurants is shown in Table 2.4. As evident from the figures, the number of stores and sales per stores grew rapidly in 1997–1998. A decline in 1980–1981 is primarily attributed to the economic recession of that period. For this segment of restaurants, 1985 was described by many as the year of new product introductions and the development of new marketing strategies. These developments were primarily due to the challenges faced by the restaurant industry and increased competitiveness within this segment. This is the oldest and the largest segment of the industry and, in many instances, is the indicator of the future trends for franchise restaurants. Competition for restaurant sites, both regionally and nationally, is growing, perhaps to saturation point. Sandwich chains, ranked by number of units by the *Nation's Restaurant News,* are presented in Table 2.5. McDonald's led the group with 12,380 units, followed by Subway and Burger King.

TABLE 2.4 SANDWICH CHAINS RANKED BY U.S. SYSTEMWIDE SALES GROWTH
(YEAR-TO-YEAR PERCENTAGE CHANGE)

Latest Year Rank	Preceding Year Rank	Chain	Fiscal Year End	Latest* vs. Preceding	Preceding vs. Prior
1	1	Schlotzsky's Deli	Dec. '97	33.49%	41.14%
2	2	Blimpie Subs & Salads	June '98	20.21	24.24
3	3	Sonic Drive-In	Aug. '97	16.03	11.84
4	14	Krystal	Dec. '97	13.01	1.39
5	5	Jack in the Box	Oct. '97	12.12	9.45
6	4	Carl's Jr.	Jan. '98	9.60	11.61
7	7	White Castle	Dec. '97	9.56	7.90
8	6	Burger King	Sept. '97	8.22	8.96
9	12	Arby's	Dec. '97	7.12	2.75
10	10	Subway	Dec. '97	6.28	3.85
11	11	McDonald's	Dec. '97	4.61	2.92
12	9	Wendy's	Dec. '97	4.58	5.37
13	8	Whataburger	Sept. '97	3.15	7.49
14	13	Dairy Queen	Nov. '97	2.04	2.08
15	15	Taco Bell	Dec. '97	1.64	−0.54
16	17	Checkers Drive-In	Dec. '97	−5.39	−11.09
17	18	Rally's Hamburgers	Dec. '97	−8.38	−12.04
18	16	Hardee's	Dec. '97	−13.26	−8.12
		AVERAGE:		6.92%	6.07%

* Actual results estimates or projections
Source: NRN Research

In order to meet consumer demands, menu changes and convenience are two factors that are being given priority by restaurants in this segment. Increasing dependence on international expansion for sustained growth is another trend followed by those companies facing domestic prime site scarcity. However, the changing financial situation overseas had a big impact in 1998. Strategy by some of the restaurant chains includes multisegment diversification, ethnic appeal, and sensitivity to service and value concerns of the customers.

TABLE 2.5 SANDWICH CHAINS RANKED BY NUMBER OF U.S. UNITS

Latest Year Rank	Preceding Year Rank	Chain	Fiscal Year End	Year-End Number of Units		
				Latest*	Preceding	Prior
1	1	McDonald's	Dec. '97	12,380	12,894	11,368
2	2	Subway	Dec. '97	11,165	10,848	10,093
3	3	Burger King	Sept. '97	7,414	6,925	6,492
4	4	Taco Bell	Dec. '97	6,768	6,670	6,498
5	5	Dairy Queen	Nov. '97	5,052	5,035	5,089
6	6	Wendy's	Dec. '97	4,575	4,369	4,197
7	7	Hardee's	Dec. '97	2,944	3,225	3,395
8	8	Arby's	Dec. '97	2,913	2,859	2,798
9	9	Blimple Subs & Salads	June '98	2,175	1,800	1,581
10	10	Sonic Drive-In	Aug. '97	1,680	1,567	1,464
11	11	Jack in the Box	Oct. '97	1,317	1,244	1,231
12	12	Carl's Jr.	Jan. '98	686	643	633
13	13	Schlotzsky's Dell	Dec. '97	653	557	463
14	14	Whataburger	Sept. '97	534	516	511
15	15	Checkers Drive-In	Dec. '97	478	478	499
16	16	Rally's Hamburgers	Dec. '97	477	444	481
17	17	Krystal	Dec. '97	349	338	336
18	17	White Castle	Dec. '97	310	305	296
		TOTAL:		61,870	59,917	57,328

* Actual results, estimates or projections
Source: NRN Research

As far as menu diversification is concerned, many sandwich restaurants have already introduced or are testing items such as chicken, pizza, pasta, regional recipes, and prepackaged salads. Also, some of the hamburger chains are interested in developing special menu items with a variety of flavor. Prepackaged salads have replaced salad bars in many restaurants due to the high labor, food, and operational costs involved in salad bars. Upkeep of the food quality and sanitary conditions are constant concerns of unit managers. As for consumer convenience, many franchised restaurants are experimenting with features such as double drive-throughs, mobile units, and home delivery. In addition to adapting to changing consumer tastes and increased competition, these chains are faced with the challenge from nutrition-conscious consumers and consumer groups. Many chains are cutting down on calories by promoting items such as low-fat shakes, yogurt, and fat-free muffins. Most of the major sandwich chains have also switched from the use of lard and beef tallow to all-vegetable oils for frying.

Discounting and special sales are becoming popular ways to combat competition within this segment of the industry. The toughest competition faced by the hamburger segment comes from the non-burger types of restaurants. The consumer tendency is to shy away from burgers. Pizza and pasta seem to be competing vigorously with traditional hamburger restaurants.

Menu diversification adds to operational problems due to increased inventories, labor costs, and quick customer service. Taco Bell introduced new items such as Gorditas, which added a different flavored product. Quality, freshness, and taste are being given importance in developing new products.

Pizza

Pizza restaurants represented the fastest-growing segment in the restaurant business in the past decade and have expanded rapidly, as evident from data shown in Table 2.6. The compounded annual growth rate was 7.5 percent. Although not as mature, the pizza restaurant segment is in strong competition with sandwich chains to the extent that many of the larger chains planned to add pizza or pizza-type menu items to their menu offerings. Chuck E. Cheese's was ranked number one in sales per unit (Table 2.7), followed by Papa John's, Round Table Pizza, Sbarro, Pizza Hut, Domino's Pizza, and Godfather's Pizza. With respect to number of units (Table 2.8), Pizza Hut is the leader with 8,698 units, followed by Domino's Pizza (4,431 units) and Little Caesar's Pizza (3,900 units). Also, as shown in Table 2.9, Pizza Hut leads the market share rankings, according to the study conducted by the *Nation's*

TABLE 2.6 PIZZA CHAINS RANKED BY U.S. SYSTEMWIDE SALES GROWTH
(YEAR-TO-YEAR PERCENTAGE CHANGE)

Latest Year Rank	Preceding Year Rank	Chain	Fiscal Year End	Latest* vs. Preceding	Preceding vs. Prior
1	1	Papa John's Pizza	Dec. '97	49.12%	34.99%
2	3	Chuck E. Cheese's	Dec. '97	9.85	6.35
3	2	Domino's Pizza	Dec. '97	7.53	9.52
4	6	Sbarro, The Italian Eatery	Dec. '97	5.38	1.37
5	5	Godfather's Pizza	May '98	2.00	2.12
6	4	Round Table Pizza	June '98	1.69	2.94
7	7	Little Caesars Pizza	Dec. '97	−1.79	−3.45
8	8	Pizza Hut	Dec. '97	−4.61	−7.04
		AVERAGE:		7.46%	5.85%

* Actual results estimates or projections
Source: NRN Research

TABLE 2.7 PIZZA CHAINS RANKED BY SALES PER UNIT

Latest Year Rank	Preceding Year Rank	Chain	Fiscal Year End	Sales Per Unit (By Fiscal Year, in Thousands)		
				Latest*	Preceding	Prior
1	1	Chuck E. Cheese's	Dec. '97	$980.0	$902.0	$827.0
2	4	Papa John's Pizza	Dec. '97	713.0	683.0	610.0
3	3	Round Table Pizza	June '98	701.0	688.2	688.2
4	2	Sbarro, The Italian Eatery	Dec. '97	693.0	698.0	702.0
5	5	Pizza Hut	Dec. '97	630.0	620.0	651.0
6	6	Domino's Pizza	Dec. '97	568.0	541.2	495.0
7	7	Godfather's Pizza	May '98	504.5	500.0	500.0
8	8	Little Caesars Pizza	Dec. '97	348.5	350.0	350.0
		AVERAGE:		$642.3	622.8	602.9

* Actual results, estimates or projections
Source: NRN Research

TABLE 2.8 PIZZA CHAINS RANKED BY NUMBER OF U.S. UNITS

Latest Year Rank	Preceding Year Rank	Chain	Fiscal Year End	Year-End Number of Units		
				Latest*	Preceding	Prior
1	1	Pizza Hut	Dec. '97	8,698	8,910	8,883
2	2	Domino's Pizza	Dec. '97	4,431	4,300	4,242
3	3	Little Caesars Pizza	Sept. '97	3,900	4,008	4,187
4	4	Papa John's Pizza	Dec. '97	1,517	1,160	878
5	5	Sbarro, The Italian Eatery	Nov. '97	805	757	725
6	6	Round Table Pizza	June '98	603	579	559
7	7	Godfather's Pizza	May '97	554	540	525
8	8	Chuck E. Cheese's	Dec. '97	312	314	315
		TOTAL:		20,820	20,568	20,314

* Actual results, estimates or projections
Source: NRN Research

TABLE 2.9 PIZZA CHAINS RANKED BY TOP 100 MARKET SHARE (SHARE OF AGGREGATE SALES OF PIZZA CHAINS IN TOP 100)

Latest Year Rank	Preceding Year Rank	Chain	Fiscal Year End	Annual Market Share		
				Latest*	Preceding	Prior
1	1	Pizza Hut	Dec. '97	43.52%	46.64%	50.05%
2	2	Domino's Pizza	Dec. '97	22.97	21.77	19.83
3	3	Little Caesars Pizza	Dec. '97	12.73	13.25	13.69
4	4	Papa John's Pizza	Dec. '97	8.03	5.86	4.33
5	5	Sbarro, The Italian Eatery	Dec. '97	3.91	3.79	3.73
6	6	Round Table Pizza	June '98	3.62	3.64	3.53
7	7	Chuck E. Cheese's	Dec. '97	2.71	2.54	2.38
8	8	Godfather's Pizza	May. '97	2.51	2.51	2.46
		TOTAL:		100.00%	100.00%	100.00%

* Actual results, estimates or projections
Source: NRN Research

Restaurant News. It has a market share of 43.5 percent, followed by Domino's Pizza (23 percent) and Little Caesar's Pizza (12.7 percent). These 1998 rankings are similar to those in 1997.

The intense competition in this segment led to a variety of concept developments. As an example, home delivery service to the consumer propelled this segment ahead of the hamburger segment as the growth leader in foodservice. Express pizza restaurants opened at shopping centers and malls. Hamburger chains are adding pizza to their menus to stay in competition with pizza restaurants. Many other restaurants are adding pizza delivery service and those doing strictly delivery business are planning to add table service restaurants. Pizza Hut is now offering home delivery service to compete in the marketplace. Some showbiz-type restaurants, such as ShowBiz Pizza and Chuck E. Cheese Pizza, are developing new plans to capture pizza-loving consumers, particularly children. Small regional pizza chains are working on franchise expansion. It is predicted that off-premises pizza dining will continue to grow in the coming decades. Carry-out and delivery units are expected to grow at a rapid pace. Delivery-only units tend to be far smaller than full-service pizza restaurants and therefore involve lower capital expenses. Product quality, improving production technology, and renovations are primary aspects of this segment.

The primary reason for promising growth by the pizza segment is its exceptionally broad demographic base—crossing all age groups, sexes, regions, and income levels—compared to other foods in the market. The suitability of this product for delivery is another major attribute of its popularity. Convenience and time saving will be the major factors considered by the computer- and technology-oriented consumers of the next century. Pizza will suit the tastes of such consumers.

Chicken

Chicken received considerable attention due to growing nutrition-consciousness of consumers. Many of the hamburger and sandwich chains added chicken to their menu, thereby intensifying the already stiff competition within this segment of the restaurant industry. Rival sandwich chains were scrambling to add chicken to their menus in addition to or in lieu of beef items. In order to counterbalance fried items, many chains are adding grilled chicken items or are in the process of test-marketing them.

Chicken chains ranked by market share are listed in Table 2.10. Kentucky Fried Chicken is on top with almost 56 percent of the annual market share, followed by Boston Market as the distant second. Chicken chains by number of U.S. units are shown in Table 2.11. Kentucky Fried Chicken leads again with 5,120 units, followed by Boston Market with 1,166 units and Churchs Chicken with 1,070 units. Popeye's, and Chick-Fil-A, follow with fewer than 1,000 units each.

Table 2.12 shows the growth of the franchised chicken segment of the restaurant industry since 1973. Growth in the number of units has fluctuated at an average of about 5 percent. As with other segments, innovative and creative changes are being sought to cope with the growing competition. Some restaurants are using boneless chicken meat in sandwiches while others are experimenting with chicken with bone. Menus keep enlarging

TABLE 2.10 CHICKEN CHAINS RANKED BY TOP 100 SHARE (SHARE OF AGGREGATE SALES OF CHICKEN CHAINS IN TOP 100)

Latest Year Rank	Preceding Year Rank	Chain	Fiscal Year End	Annual Market Share		
				Latest*	Preceding	Prior
1	1	KFC	Dec. '97	56.07%	57.65%	60.52%
2	2	Boston Market	Dec. '97	16.78	16.26	12.35
3	3	Popeyes	Dec. '97	10.09	9.85	10.73
4	4	Chick-fil-A	Dec. '97	9.02	8.43	8.20
5	5	Churchs Chicken	Dec. '97	8.04	7.82	8.19
		TOTAL:		100.00%	100.00%	100.00%

* Actual results, estimates or projections
Source: NRN Research

TABLE 2.11 CHICKEN CHAINS RANKED BY NUMBER OF U.S. UNITS

Latest Year Rank	Preceding Year Rank	Chain	Fiscal Year End	Year-End Number of Units		
				Latest*	Preceding	Prior
1	1	KFC	Dec. '97	5,120	5,079	5,152
2	2	Boston Market	Dec. '97	1,166	1,087	829
3	3	Churchs Chicken	Dec. '97	1,070	989	964
4	4	Popeyes	Dec. '97	945	892	884
5	5	Chick-fil-A	Dec. '97	749	717	657
		TOTAL:		9,050	8,764	8,486

* Actual results, estimates or projections
Source: NRN Research

with different types of chicken, and there is evidence of constant test-marketing by many leading chains. For example, Chick-Fil-A's menu includes chicken nuggets, barbecued chicken sandwiches, broiled chicken sandwiches, chicken salad sandwiches, chicken salad plates, and breast of chicken soup. Nonfried items are gaining popularity over fried. Mexican-style grilled chicken and spicy chicken are also being test-marketed. This segment will face an

TABLE 2.12 CHICKEN CHAINS RANKED BY GROWTH IN NUMBER OF U.S. UNITS (YEAR-TO-YEAR PERCENTAGE CHANGE)

Latest Year Rank	Preceding Year Rank	Chain	Fiscal Year End	Latest* vs. Preceding	Preceding vs. Prior
1	3	Churchs Chicken	Dec. '97	8.19%	2.59%
2	1	Boston Market	Dec. '97	7.27	31.12
3	4	Popeyes	Dec. '97	5.94	0.90
4	2	Chick-fil-A	Dec. '97	4.46	9.13
5	5	KFC	Dec. '97	0.81	–1.42
		AVERAGE:		5.33%	8.47%

* Actual results estimates or projections
Source: NRN Research

increasing challenge, as almost every restaurant segment is planning to add chicken to the menu.

Discussion of the preceding segments illustrates the variety of chain restaurants. Other segments of the franchised restaurants include dinner houses, cafeterias, and family restaurants, but are not discussed in detail, since the intent was to focus on major segments that are closely involved in franchising. Also, these segments were focused as examples.

Restaurant Study 2

A&W Restaurants, Inc.

One hot day in June 1919, in Lodi, California, an entrepreneur named Roy Allen mixed up a batch of creamy root beer and sold the first frosty mug of this delightful beverage for one nickel. Now, 70 years later, A&W Root Beer is the world's number-one selling root beer and is still made fresh daily and sold at hundreds of A&W Restaurants.

Allen purchased the formula for his root beer from a pharmacist in Arizona. To this day, the unique blend of herbs, spices, barks and berries remains a proprietary secret.

With the success of his first root beer stand in Lodi, Allen soon opened a second stand in nearby Sacramento. It was there that what is thought to be the country's first "drive-in" featuring "tray-boys" for curb side service, opened up.

In 1922, Allen took on a partner, Frank Wright, an employee from his original Lodi location. The two partners combined their initials—"A" for Allen and "W" for Wright—and formally named the beverage, A&W Root Beer. Three units were opened in Sacramento, then on to other northern and central California locations and to the states of Texas and Utah.

In 1924, Allen bought Wright's share of the business to actively pursue a franchise sales program. He had the name, A&W Root Beer and the A&W logo legally trademarked with the U.S. Patent and Trademark office.

By 1933, the creamy beverage was such a success that Allen had more than 170 franchised outlets operating in the Midwest and West. To insure uniform quality for the namesake beverage, Allen sold A&W Root Beer concentrate exclusively to each franchise operator. His profits were derived from the sale of the concentrate and a nominal license fee.

During World War II no new restaurants were opened and despite governmental sugar rationing (this affected supplies of bottler's sugar, a necessary ingredient of root beer) and employee shortages (also a result of the war) most A&W units remained successful. After the war, the number of A&W restaurants tripled as GI loans paved the way for private enterprise to flourish.

In 1950, with more than 450 A&W's operating nationwide, founder Roy Allen retired and sold the business to an aggressive Nebraskan named Gene Hurtz who formed the A&W Root Beer Company. The postwar era—the rapidly recovering economy and popularity of the automobile, provided the right environment for Hurtz's company to prosper. Drive-in's were becoming increasingly popular and A&W had the privilege of being one of the few nationally established drive-in restaurant chains. By 1960 the number of A&W's had swelled to over 2,000.

The first A&W restaurants outside of the United States opened in 1956 in Winnipeg, Manitoba, Canada (the Canadian division eventually became a wholly owned subsidiary of A&W and in 1972, was sold to Lever Brothers, Ltd., an international conglomerate).

In 1963, the A&W Root Beer Company was sold to the J. Hungerford Smith Company, the firm that had manufactured A&W Root Beer concentrate since 1921.

In that same year, the first overseas A&W Restaurant opened its doors. Located in Guam, the international division quickly expanded to the Philippines.

Three years later, both A&W and J. Hungerford Smith Company were purchased by United Fruit Company of Boston. In 1970, United Fruit was acquired by The AMK Corporation, which formed the new corporation, United Brands Company. Within this structure the A&W Root Beer Company adopted a new trademark, changed its name to A&W International, Inc. and began the process of becoming a full-fledged restaurant and food service organization.

Many innovative changes were instituted. One was the formation of the National Advisory Council of the National A&W Franchisees Association (NAWFA). This elected board marked the first time in fast-food industry history that franchisees had a voice in the formation of their contract, a move Ralph Nader lauded as being one of the fairest in the industry.

The new contract featured a revised royalty agreement. While corporate profits would continue to benefit from the sale of new franchises, other revenue would no longer be generated from the sale of beverage equipment and food items. The new agreement held that each franchisee's royalty fees would be based on a percentage of their restaurant sales. The corporation was now profitable in direct relation to the success of the chain.

Other changes included the expansion of a nationwide distribution network allowing franchisees to purchase concentrates, food items, paper goods, and glass mugs. And, programs offered by the corporation were revamped to suit franchise needs—training, marketing, accounting, product development, bookkeeping systems, building design, and equipment layout.

America loved the taste of A&W Root Beer. So, in 1971, United

Brands formed a wholly-owned subsidiary, A&W Beverages, Inc., for the purpose of making A&W Root Beer available on the grocery shelf. First introduced in Arizona and California, the cans and bottles of A&W Root Beer were an instant success. Retailers nationwide were soon carrying the product.

In 1974, A&W Beverages, Inc. introduced A&W Sugar-free Root Beer and their goodwill ambassador, The Great Root Bear. This life-size, lovable mascot has been charming children and adults at grand openings, parades, fairs, and community visits ever since.

A standard core menu for the restaurants was introduced in 1978. It was the first time in A&W history that there was a consistent menu offering. And, it was at this time that A&W Restaurants, Inc., the wholly-owned restaurant franchise subsidiary was formed.

The corporation launched a new restaurant concept in 1978, The A&W Great Food Restaurant. A modern upscale concept, these facilities featured fresh one-third and one-half pound 100 percent pure beef hamburgers, salad bars, ice cream bars, and of course, A&W Root Beer in a frosty mug. This concept was perhaps ahead of its time, and while they still exist, they have been reformatted to blend in with the current chainwide concept—a modern, comfortable fast-food environment at competitive prices serving the finest quality food.

A. Alfred Taubman, a developer of shopping centers and real estate, purchased A&W Restaurants, Inc. in 1982, and a new era for A&W had begun.

A period of reorganization and planning began. Franchising efforts were halted while a new prototype was being developed, and new menu concepts and management techniques were being implemented. In September 1986, E. Dale Mulder, a multiple-unit A&W franchise owner and former Executive Director of the National A&W Franchise Association, was appointed President of the corporation.

The international division of A&W Restaurants has expanded its operations into seven Southeast Asian countries. A corporate office located in Malaysia is the home of A&W's international operations.

Since Mulder's appointment, franchising has begun again and a steady growth plan implemented. Emphasis has been placed on serving franchise needs and providing assistance to the operators. Today, A&W Restaurants number over 530 both domestically and internationally.

Looking ahead to the 1990s, A&W Restaurants' future is bright. A recent agreement to convert more than 200 Carousel Snack Bars of Minnesota to A&W Hot Dogs and More will spark what is bound to be an exciting period in the history of this chain that is known for A Taste To Remember, its frosty mug of A&W Root Beer and good food.

In January 1991, the momentum of the chain increased greatly with the addition of George E. Michel as A&W Restaurant's new President and Chief Operating Officer. A 20-year veteran of A&W Food Services of Canada, Mr. Michel is credited with more than doubling the number of corporate restaurants during his tenure in Canada and is also recognized for guiding the company's growth by increasing the number and strength of the franchise operations.

Mr. Michel guided A&W Restaurants' expansion efforts into the captive, high-pedestrian segment with an immediate focus on opening new restaurants in food courts, shopping centers, office buildings, and multiuse complexes. This exciting direction is underway both corporately and in the A&W franchise community.

MILESTONES IN THE HISTORY OF A&W RESTAURANTS, INC.

June 20, 1919	The nation's root beer industry begins when Roy Allen opens the first root beer stand in Lodi, California, to sell a proprietary beverage made from a secret blend of 16 herbs, spices, barks, and berries.
1920	Allen's enterprise doubles with the opening of a second outlet in Sacramento.
1922	Allen goes into partnership with one of his employees, Frank Wright, and three new outlets are established in Houston. The partners give their root beer the famous A&W name.
1923	Two more outlets open in Sacramento, where Allen develops the nation's first "car hop" restaurant service.
1924	Allen buys out Wright's interest in the company, and registers A&W as the formal legal trademark for his product.
1925	Allen begins selling franchises to others, enabling them to open A&W drive-ins and vending booths, thus establishing America's first franchise restaurant chain.
1933	In the course of eight years, Allen has sold 171 franchises nationwide, with heavy concentration in the midwestern states.

1941–45	No new outlets are opened during World War II, primarily because of sugar and manpower shortages.
1945–50	In the postwar boom years. Allen's franchises become a hot item, particularly for veterans who want to open their own businesses. By 1950, some 450 A&W drive-ins—many featuring food—are opened.
1950	Roy Allen retires, selling his business to Gene Hurtz of Nebraska.
1956	A&W goes international, opening a drive-in in Winnipeg, Manitoba, and begins an aggressive franchising program in Canada. (The Canadian division became a wholly-owned subsidiary and in 1982 was sold to Lever Bros., Ltd.)
1962	The first overseas A&W Restaurant opens in Guam. The international division quickly expanded to the countries of the Philippines and Malaysia.
1963	The A&W Root Beer Company is sold to J. Hungerford Smith Company.
1966	United Fruit Company of Boston purchases J. Hungerford Smith Company. (United Fruit becomes United Brands Company in 1970.)
1968	Founder Roy Allen dies. A&W Root Beer Company changes its name to A&W International.
1970	A new royalty franchise system is negotiated with the National Advisory Council of the National A&W Franchisees Association. Franchisees are now able to purchase syrup, food supplies, paper goods, and glassware through A&W's central purchasing and distribution system.
1971	United Brands forms A&W Beverages, Inc. to produce cans and bottles of A&W Root Beer. These were first introduced in San Diego, California, and Phoenix, Arizona, and due to a successful introduction, a rapid national rollout began.
1972	More than 80 percent of A&W's franchisees have signed the new royalty franchise contracts.
1974	A&W's Goodwill Ambassador mascot, the Great Root Bear is created to participate in grand openings and perform community services, such as entertaining at children's hospitals. Sugar-Free A&W Root Beer was introduced.

1976	Just five years after the introduction of A&W Root Beer in cans and bottles, it becomes the nation's leading root beer.
1978	All franchisees are given a national "core menu" to standardize their food offerings. The restaurant chain's name is changed to A&W Restaurants, Inc.
1979	A new restaurant concept emerges with the opening of an A&W Great Food Restaurant in Gaithersburg, Maryland. The concept has been specially designed to offer customers quality dining with prompt service while retaining the traditional values that have been part of A&W from its beginnings.
1982	A. Alfred Taubman purchases A&W Restaurants, Inc. from United Brands Company, and corporate offices are transferred from Santa Monica, California, to Michigan. Franchising is temporarily halted during a reorganization and planning period.
1983	United Brands sells A&W Beverages (the can and bottle company) to a company now known as A&W Brands.
1986	A grilled chicken sandwich which is prepared in a more healthful way is introduced to meet the consumer's demand for healthier, lower calorie fast-food selections. A&W Restaurants is the first in the industry to offer this marinated, grilled breast of chicken served on a cracked-wheat bun. Dale Mulder is appointed President of A&W Restaurants, Inc. Mulder has been an A&W Restaurant franchisee since 1961 and former Executive Director of the National A&W Franchise Association. Emphasis is placed on updating existing restaurants. An aggressive remodel program and franchise sales campaign begins after a period of reorganization which began in 1982.
1988	A&W Restaurants launch an exciting new product—Curly Fries. These lightly battered, spicy potatoes are shaped like a corkscrew and cooked in all-vegetable oil. First drive-thru only unit opens in Flint, Michigan. This drive-thru only unit is one of several prototype

	designs A&W Restaurants is recommending to new and existing franchisees.
1989	Carousel Snack Bars, Inc. becomes a master licensee of A&W, with plans to convert over 200 existing mall and food court units to A&W Hot Dogs and More by the year 1991.
	Cholesterol-free Curly Fries become the standard for the chain, replacing the straight cut fries. This includes the switch to all-vegetable shortening for operators.
	The company has stabilized and growth is occurring at a steady pace. Since franchising began once again in 1986 more than 60 new domestic and international units have opened their doors.
1991	George Michel is appointed President and Chief Operating Officer, with Dale Mulder becoming Chief Executive Officer. Michel joins the A&W Restaurant chain with 20 years of diverse experience from A&W Food Services of Canada, Ltd.
	Mr. Michel worked closely with Dale Mulder, Chief Executive Officer, as they focused the Company's expansion efforts in the captive, high pedestrian segment. The immediate focus was aimed at opening new restaurants in food court locations in addition to office buildings, multiuse complexes, corporate cafeterias, and shopping centers.
	This exciting new direction, underway both corporately and in the A&W franchise community, with 28 food courts currently operating and approximately 15 new locations projected to open in 1992.
1992	In January, George Michel became Chief Executive Officer of A&W Restaurants, Inc. and continues to strengthen and lead the Company toward targeted growth.
1993	Cadbury Beverages acquires A&W Brands.
1994	A&W Root Beer turns 75 years old. This milestone is celebrated jointly by A&W Restaurants, Inc. and by Cadbury Beverages. *Nation's Restaurant News* lists A&W as the nation's oldest restaurant chain.
1995	A&W Restaurant Acquisition Company, Inc., headed by Sidney Feltenstein, purchases A&W Restaurants, Inc. from Taubman interests. The new ownership

backed by the investment company, Grotech Capital, will draw upon Mr. Feltenstein's long history of industry experience (he is a former Executive Vice President of Marketing for Burger King Worldwide and a onetime key executive with Dunkin' Donuts) to help move A&W Restaurants into a new era of growth and prosperity as it nears the year 2000.

(a)

(b)

An exterior view of a typical A&W Restaurant (courtesy A&W Restaurants, Inc.)

Restaurant Study 3

Golden Corral Steaks, Buffet & Bakery

Date Established .... January 1973
Company Owned Units (as of 8/31/98) 155
Franchised Units (as of 8/31/98) 293
Total Units (as of 8/31/98) 448
Registered States .... All
Cash Investment .... $250,000 Minimum
Total Investment .... $1,229,200–$3,340,000
Fees:
    Franchise $40,000
    Royalty 4%
    Advertising .... 2% Local Minimum
Contract Period(s) .... 15 years
    plus two 5-year options
Area Development Agreement .... Yes
Financial Assistance Provided .... 3rd party
Site Selection Assistance .... Yes
Expansion Plans .... U.S. Canada,

Golden Corral family restaurants feature "steaks, buffets, and bakery." The "Golden Choice Buffet" offers 140 hot and cold items. A special feature is the "Brass Bell Bakery" which prepares made-from-scratch rolls, cookies, muffins, brownies, and pizza. Steak, chicken, and fish entrees are also available. Value-driven concept with a $6 per person check average. Open lunch and dinner 7 days—weekend breakfast buffet.

General/Training is provided by an industry-leading 12-week training program. First 10 weeks are in certified training restaurants with final 2 weeks in corporate center for training and development.

Expansion throughout the U.S., Canada, and Puerto Rico is planned. Fastest growing family restaurant chain in the U.S. New store opening every five days.

Three different building sizes offered to tailor the land and building investment to the population and demographics of the market. There is an active franchise advisory council. Golden Corral is a privately-held company based in Raleigh, North Carolina.

(a)

(b)

One of the buildings offered by Golden Corral Steaks, Buffet & Bakery Restaurant (courtesy Golden Corral)

CHAPTER 3

Pros and Cons of Franchising

Franchising presents an exciting opportunity for both franchisor and franchisee, who share a symbiotic relationship. Much of the success of any franchise system depends on how smooth this relationship is. Because both franchisors and franchisees have personal goals that can be achieved by mutual cooperation and trust, it is imperative that they understand all operational aspects of the business. Many franchisor-franchisee relationship problems can be traced to informational imbalance, which can occur before or after entering into contract. Some franchisees do not understand the fundamentals of franchising and at times are impressed with the apparent glamour of successful franchise businesses, which they may observe without being directly involved. The intent of this chapter is to outline the pros and cons of franchising, viewed from the point of view of franchisees as well as franchisors. A summary of advantages and disadvantages is presented in Table 3.1.

Advantages to Franchisees

Several advantages to franchisees result from their participation in a franchise system. The most important ones are discussed below.

Established Concept

The franchisee buys into a business that has an established concept. The product or service provided is unique and has a potential for success. In business-format franchising, which includes most franchised restaurants, the entire format is available for use. If the franchise has been in operation for some time, consumers are aware of the company, and in many cases the reputation of the product or service is well established. Thus a franchisee is

TABLE 3.1 ADVANTAGES AND DISADVANTAGES OF FRANCHISING

Advantages	Disadvantages
To Franchisee	
Established concept	Unfulfilled expectations
Tools for success	Lack of freedom
Technical and managerial help	Advertisement and promotion practices
Standards and quality control	Cost of services
Minimum risk	Overdependence
Less operating capital	Monotony and lack of challenge
Access to credit	Termination and renewal
Comparative assessment	Other problems
Research and development	
Advertisement and promotion	
Other opportunities	
To Franchisors	
Business expansion	Lack of freedom
Buying power	Franchisee's financial situation
Operational convenience	Franchisee recruitment, selection, and retention
Franchisee's contribution	Communication
Motivation and cooperation	

buying into an established business—a major advantage. In the case of restaurants, if the menu is well known and consumers are aware of the trademark, the business is free from start-up uncertainties. Established franchise restaurants have undergone years of rigorous market testing so that the concept is so well proven that it is ripe for profitable applicability. Although this advantage is available to the franchisee at the time of initiating the business, further success depends on the way the individual franchise is operated and managed.

Tools for Success

Franchising does not guarantee success but does provide tools for success. These tools include: (a) local and national support by the franchisors in vital areas such as site selection, construction, purchasing, equipment selection, operations, training, advertising, marketing, and promotion, and (b) availability of continuing assistance in various facets of the business from the regional or corporate offices of the franchisors.

Many franchisors provide turnkey operations to their franchisees, which avoids the problems associated with starting a traditional nonfranchised restaurant operation. The primary reasons why the restaurant business is so challenging are that it is complex and multifaceted. It involves expertise from different disciplines. An individual alone finds it difficult and cost-prohibitive to afford these tools. Franchisors, due to centralization of collective resources, can provide such services to franchisees. In addition to all these benefits, the time and effort needed to establish a new business are lower in joining a franchise system than in starting a restaurant from scratch.

Technical, Operational, and Managerial Assistance

Major advantages of franchising include the technical and managerial assistance provided by the franchisor. This assistance allows even an inexperienced person to enter a business with which he or she is unfamiliar. Assistance is provided prior to and after opening of a business. The ongoing assistance program helps in the day-to-day operation of the business as well as during crises. Ideally, a communication line is established between franchisor and franchisees with mutually beneficial exchanges.

Several technical and operational areas may need help in any business venture. Some areas of technical assistance provided by restaurant franchisors include market feasibility studies, site selection, architectural layout and design, and equipment selection and layout. Operational assistance provided by the franchisor includes inventory purchase and control, purchase specifications, production guidelines, operation schedule, sanitation control, and other service parameters. In addition, training and operational manuals provide a continuous source of guidance. Such services make going into business easier for franchisees, although not all franchisors provide all of them.

Standards and Quality Control

Advantages of franchising include the availability of standards set by the franchisor and a mechanism for maintaining quality control. It is imperative for the success of individual franchising units that they follow fixed standards for control of the quality of products and services. Established standards are the primary reason why a hamburger sold in a franchise in California tastes and looks exactly like one sold by the same franchise in New York. The same is true for the decor, overall theme, and services provided by a franchise restaurant. Mutual cooperation and effective administration are essential for the maintenance of uniformity of the product and services. Uniformity is essential to build and retain the image, to ensure return business, and to maintain employee morale and the growth and development of the enterprise. If properly followed, standards also help assure teamwork between the franchisor and the franchisees and their employees.

Implementation of standards require that they be reasonable, flexible, and applicable to specific areas.

Minimum Risk

There is no business without risk—risk being the potential for both success and failure. Franchising reduces the risk of failure to a great extent. The risks involved are substantially lower than when starting a business from scratch. Because franchising, on the basis of research and expertise, leads to the development of a system that has been proven profitable and workable, there is minimum risk of failure. To be exact, franchising offers risk reduction rather than risk elimination. Franchising does not guarantee success, and each unit within a franchise faces the possibility of risk. Much depends on how efficiently a franchisee manages business. It is rightly said that when the relationship between the franchisor and the franchisee is set up and operated properly, the result is as close to assured business success as one can get.

Less Operating Capital

Compared to independent restaurants, franchise restaurants may require less operating capital. In a franchise restaurant, certain items, like inventory level, can be forecast more accurately than in an independent unit. Calculated production and portion control methods reduce the chances of wastage or unnecessary maintenance of stocks. Expenses can be reduced similarly in other aspects of management. Financial credits in the form of inventory or equipment may also be obtained from the franchisor. Ingredients and supplies (particularly during the initial stages of the operation) may be provided on credit by the franchisor. The design of the facilities, well planned by the franchisor's experienced staff, results in increased production and efficient service.

Franchisors may also help indirectly by facilitating the purchase of business insurance, employee health insurance, and other benefit packages. The group buying power and affiliation with insurance companies by the franchisors may help franchisees reduce their operating costs.

Access to Credit

A major advantage of franchising is the access to financial credit provided by the franchisors. Although this may not apply to a new franchisee, often franchisors provide credit to franchisees for business expansion on much more favorable terms than other financial institutions. This in-house support is looked on favorably by many franchisors as well, as it indirectly has a positive impact on their business. Thus there is mutual benefit in this sort of assistance.

Comparative Assessment

The comparative assessment of functional and operational activities is facilitated within a restaurant franchise system. Uniformity of operation, unique to franchise systems, facilitates the comparison of one restaurant unit with another and consequent critical evaluation of the business at various levels. Franchisors can provide information pertaining to the successes of and new approaches developed by other franchisees within the system. Meetings with other franchisees also help in this evaluation. There are shared benefits in learning about the experiences of business counterparts.

Benefits from Research and Development

Franchisors have an ongoing interest in the development of their franchises, and many maintain permanent research and development departments. The fruits of these efforts are shared with the franchisees. Research can be in the area of restaurant products or services. This type of service is not available to most independents or nonfranchised restaurants. Further, acute individual problems can be referred by the franchisees to these departments for solutions.

Advertising and Promotion

Funds pooled from all franchisees are available for advertising and promotion—thus, wider exposure can be attained. Franchisees have combined buying power because of the pooling of resources and possibly by centralized operations. Carefully planned advertising done on a large scale by qualified staff and agencies can be a real asset for a franchisee. Similarly, well-researched promotional efforts can enhance the profitability of an operation.

Opportunities

For a franchisee, unique opportunities are provided by franchising. The opportunity to own one's own business and the rewards associated with success are both provided by successful franchises. There is a chance to be in business for oneself without having a business by oneself. Independence is intertwined with mutual dependence, often referred to as a win-win situation. Such limited autonomy is desired by many individuals. When franchising works properly, it offers access not only to the trademark and operational know-how of a large company but also its clout and expertise while still allowing the franchisee to maintain entrepreneurial independence.

There is a personal satisfaction and pride in being associated with a successful franchise. Franchisees become a part of the overall success attained by the larger, probably well-known, and profitable franchise system. Thus an opportunity to contribute to the success of the parent corporation is available to franchisees.

Franchising also provides an opportunity for personal growth and advancement in business knowledge. This may result from the extensive training, seminars, meetings, continuing education, and operational information provided by the franchisor. Hands-on experience can lead to further personal growth and advancement. The opportunity to meet other franchisees within the franchise system can lead to the exchange of ideas and a comparative assessment of the operational status of the franchise.

The opportunity to work with people and for people is considered as an asset by many franchisees. People include customers, restaurant crews, community groups, vendors, and purveyors. Further business growth opportunities are provided by the territorial rights that may be granted by the franchisors. Expansion can result from buying, subfranchising, leasing, operating, or converting other restaurants into franchises. The territorial right is considered a major benefit of franchising, one that has led to the success of franchises for decades.

The above-listed advantages are common to most franchises; however, none can be guaranteed. In short, nevertheless, franchising helps franchisees to remain competitive in business compared to independent or nonfranchised operators.

Disadvantages to Franchisees

Franchising agreements are, ideally, designed to foster the smooth relationship between franchisor and franchisee on which so much depends. The disadvantages to franchising primarily arise from an uneven franchisor-franchisee relationship that leads to dissatisfaction in either or both of the parties.

Unfulfilled Expectations

Franchisees build certain expectations before getting into the business. Franchisors at times present an unrealistic business picture. Contract clauses may be interpreted differently, building false expectations. All of these lead to dissatisfaction among the franchisees. Further, misleading or fraudulent practices of some franchisors has victimized potential franchisees.

Some franchisees fail to read or understand the implications of some of the contract clauses and rely on the sales or promotional literature provided by the franchisors. Franchisees also fail to understand the upper hand that most of the franchisors enjoy in a franchise system. A disadvantage may be rooted in the disparity of bargaining power that often occurs in franchising. Because franchisors are financially strong, legal battles often prove to be long and expensive for individual franchisees.

Lack of Freedom

In spite of several advantages, franchising poses certain severe restrictions and limitations. Franchising contracts may contain restrictions on territorial expansion or may limit potential customer contract. Territorial rights may be inequitably distributed or a territorial overlap may interfere with the prosperity of a franchisee's business. Also, an entrepreneur will have ideas that may be hindered by the restrictions imposed by a franchisor. For example, an enterprising foodservice manager may have proven creative menu ideas for regional consumers that cannot be implemented due to the franchisor's restrictions. A franchisee may find food delivery a profitable venture but may be prevented from providing the service by the policy of the franchisor.

Advertising and Promotion Practices

Although advertising and promotion are listed as advantages of franchising, under certain circumstances they can prove to be a disadvantage. A franchisee may be paying fees for advertising that is impractical or not applicable to its local market conditions. At times it may not even reach specific consumers within the territorial expanse of the franchisee, or advertising methods used may not be appropriate for the desired clientele. In other words, a franchisee may be paying for advertising directed toward other franchisees. Similarly, promotional methods selected by the franchisor may not be applicable or profitable for a franchisee's operation. For example, coupon sales promoted by franchisor may not be necessary for a restaurant, and selling discounted items, even if sold in large quantities, imposes a burden on the franchisees.

Services Provided by the Franchisor

Franchisees have to pay service costs and, if the services provided by the franchisors are not up to par, the financial loss may be considerable. Also, with experience, franchisees may come to feel, in light of the services provided in return, that the franchise fees and royalty fees are not justifiable. It may be psychologically difficult for a franchisee to accept profit sharing in the form of a fixed percentage. Franchise fees and royalty fees have an adverse impact on the return on investment of franchisees and are always sore points in the franchisee-franchisor relationship.

Overdependence

In a franchise system, a franchisee may become overdependent on the advice of franchisors in areas such as operations, crisis situations, pricing strategy, and promotions. In addition to slowing down the decision-making process, overdependence may become too costly for a franchisee. Local franchisees can make better decisions than the franchisor in some situations. Franchisees may also rely completely on promotional practices of the franchisor. Similarly, on management aspects, a franchisee may rely too heavily on the judgment of the franchisor.

Monotony and Lack of Challenge

After a certain period of time, franchisees may find business to be monotonous. Particularly for an entrepreneur, a franchise system may become too routine and lacking in challenge and scope for creativity. It may discourage an enterprising franchisee by not providing opportunities for advancement. Such franchisees may diversify their business investment, perhaps toward other types of businesses or competing franchises. This practice may lead to a lack of responsibility and individual decision making by an otherwise capable franchisee.

Franchise Termination, Renewals, and Transfers

Franchise termination is the single most important issue for franchisees, whose capital investment, years of service, and livelihood in a franchise are dependent on franchisors. The franchisor's power to terminate, to decline to renew, or to deny the franchisee the right to sell or transfer a franchise has always been a sore point in the franchisor-franchisee relationship. According to a report submitted to the 101st U.S. Congress by the Congressional Committee on Small Business:

> In order to avoid the implicit "threat" of termination, franchisees feel compelled to comply with all directives and instructions of the franchisor, no matter how unreasonable, invalid or arbitrary they may be. Most prevalent forms of abuse in this regard involve: required purchases of inventory and equipment from the franchisor at above market prices; required testing of unproven products without allowance for potential loss; required investment to alter the design or appearance of the franchise location; "voluntary" contributions for special promotional campaigns; alteration of exclusive marketing or territorial rights; and extension of non-competition agreements to apply to franchise operations or to unrelated business activities of franchisees.

Franchisors feel that such power is essential to the efficient operation of a franchise system. Thus, a franchisee's right to a franchise may be revoked if any of the provisions of the franchise agreement are not followed. This is considered as a major disadvantage by many franchisees, particularly those who have stayed with a system for a considerable period of time or who have invested substantially in a franchise system.

Franchisee's and Franchise System Performance

A franchisee's success is dependent on the performance of other franchisees within the system. If the quality and service of a franchisee are not up to the standard, the resulting deficit may have a negative impact on the entire franchise system and thereby affect the sales of other franchisees. Often, consumers blame the entire franchise system for negligence rather than an individual franchisee. This may also occur if the franchisor is not particular in maintaining quality standards uniformly among all franchisees. For example, an outbreak of a food-borne illness at one unit may have an impact on all franchise stores within that system. Thus, the performance of individual franchisees has an impact on the success or failure of the entire franchise system.

Similarly, much depends on the overall performance of the franchise system. Improper management or abrupt changes in management may affect all franchisees within a network. At times, poor performance of the franchisor is interpreted by consumers as poor performance of the individual franchisee, irrespective of how profitable a restaurant has been in the past. Another manifestation of this adverse relationship may be an outcome of the national ranking by industry publications and financial analysts. Poor rating or ranking of a franchise system may have a direct impact on sales of individual units. A guilt-by-association phenomenon may occur in such situations.

It should be clearly understood that this is an exhaustive list of the possible advantages and disadvantages of franchising and it is impossible to expect that all of these will be relevant to a particular franchise system. Also, the advantages of franchising do not necessarily mean profitability and disadvantage may not mean lack of profitability of a franchise. However, it is evident that the advantages far outweigh the disadvantages of franchising. The pros and cons of franchising, from the franchisee's point of view, are given in Table 3.1.

Advantages to Franchisors

In addition to being a financial asset, franchising provides advantages and disadvantages to franchisors, which are described in the following paragraphs.

Business Expansion

Franchising provides means for expansion of a business by providing opportunities to franchisees. This expansion is primarily provided to franchisees in territorial expansion agreements. Capital investment for such expansion can also be provided by franchising. In fact, franchising is the best way to expand with limited capital investment by the franchisor. Sometimes the excess capital is put to work in other types of investment, bringing additional profitability to the corporation. Investors are more likely to buy into an existing franchise system and help in its expansion. Also, as franchising does not involve elaborate experience of the business, many investors may join in as a restaurant franchise expands. Franchising can therefore attract capital for expansion by either direct investment by the investors or the sale of franchises. Also, potential franchisees may be located in regions with which franchisors are not familiar. A local franchisee is normally more familiar with the community, private and public organizations, zoning ordinances, license requirements, and business regulations in a region. This knowledge become an asset for the expansion of franchises.

Expansion of any corporation involves risks and structural modifications that may become difficult to handle. In franchising, the expansion takes place without any changes to the organizational structure of the corporate headquarters. This allows the franchisor more time and effort to devote to strategic planning, operational planning, market feasibility, and the overall development of the franchise system. Thus, to a franchisor there is limited risk and limited use of capital, as well as chances of equity investment.

Buying Power

Restaurant businesses involve large-quantity purchases of ingredients, equipment, and supplies. Franchisors may benefit from collective and centralized buying for franchisees. Bulk quantities of supplies or equipment can be purchased, stored, and supplied to franchisees.

In addition, funds available from franchise fees may be used for advertising, promotion, and research development on a well-planned and organized basis. Considerable cost savings can also be achieved in such ventures.

Operational Convenience

From the franchisor's point of view, franchising provides for convenient operational management compared to other businesses. A franchisor does not have to worry about employee turnover, benefits, and wages at the unit level, where such matters are handled by franchisees. Individual franchise units are more easily managed by franchisees than the franchisor's managing the entire franchise system would be. Some chain restaurants operated solely by the company face acute human resources problems. Also, as franchisees have a vested interest in their unit(s), they handle operation themselves, except under emergency or critical situations.

Franchisee Contribution

An advantage that is often underappreciated is the contribution that franchisees make to a franchise system. Large franchise corporations realize that a franchisee can make valuable contributions at the grassroots level. Franchisees are directly involved in day-to-day unit operations and thoroughly understand its functional aspects. They may provide good solutions to problems, may comment on the applicability of an idea, and may give input on the financial feasibility of a plan. When franchisees are involved, many problems can be avoided. Also, whenever a new idea or change is desired, franchisees serve as a good sounding board. Several corporate staff have repented making decisions without prior consultation with the franchisees. The best franchisors see their franchisees as tremendous assets, bringing creativity, years of hands-on experience, and the drive for success that is invaluable for a franchise system.

Many ideas originate with franchisees and, if carefully tested and adapted, may prove to be of lifetime benefit for the franchisors. As an example, McDonald's Big Mac®, Filet-O-Fish®, and Egg McMuffin® sandwiches are all reported to be based on ideas generated by franchisees. Even Ronald McDonald was created by franchisees to cultivate the children's market.

Motivation and Cooperation

Franchisees are motivated and have a vested personal interest in the success of the operation that may be nonexistent in a company-employed manager. The self-direction and motivation of franchisees have led to the success of many franchises. Also, the collective motivation and cooperative power of the franchisees behind the franchisors provide extra clout in many areas of business and regulation. However, this relationship has to be handled very carefully.

Some corporations have franchisee advisory boards that provide regional and national forums for the exchange of ideas between the company and the franchisees. This interaction has positive impact on the business health of a franchise system. Thus, these advisory boards create a check and balance system that enables the corporate functional support groups to keep a proper perspective on the franchise business.

DISADVANTAGES TO FRANCHISORS

As in the case of franchisees, there are certain disadvantages of franchising to the franchisor. Many of these disadvantages may be traced to the relationship between the franchisors and the franchisees.

Lack of Freedom

Lack of freedom on the part of the franchisors is one of the major disadvantages of franchising. There is no direct control over the units by the franchisors, which makes it difficult to make changes in policies and procedures. Also, some legal problems and suits by franchisees may hinder, delay, or adversely affect the growth and development of the franchise system. Franchisee advisory committees may become strong enough to interfere in the freelance operation of a franchise system. It might become difficult for a franchisor to modify products or processes without the cooperation of the franchisees, who may not be willing to implement a change involving their time, effort, and capital, although that change may be beneficial for the corporation on a long-term basis. Often, franchisees are interested in the immediate future and a quick return on their investment. Uncooperative franchisees can become a continuous source of problems for a franchisor.

Franchisees' Financial Situation

Franchisees' financial status, which is not under the control of the franchisors, may have an impact on a franchise system. Franchisees, particularly those who own multiple units, may declare bankruptcy, which may jeopardize the operations and overall profitability of franchisor's corporation. Also, franchisees may decide to diversify their investment and spread out so thin that their units are adversely affected.

Franchisee Selection, Recruitment, and Retention

Franchisee selection and recruitment can become a difficult task and franchisors have to be extremely careful in their selection. It is hard to come up with a profile of a successful franchisee. The glamour of franchising may attract several absentee investors who are not interested in the operational aspects of a franchise but rather in a good tax shelter. Many lack the motivation that is needed to be successful in a franchise restaurant business. Applicants may not realize the time, work, responsibility, and risks involved in franchising. Such franchisees, if allowed into a franchise system, can generate real headaches for the franchisor. Franchising is a long-term contract that binds both franchisors and franchisees. Some franchisees lose interest after recovering their initial investment or due to the monotony of activities. This adversely affects the profitability of the franchisor. Selection of franchisees is like selecting good employees: while it may be difficult to recruit them, it may be equally difficult to retain good ones.

Communication

Many franchisor-franchisee relationship problems can be traced to the communications between them. Often misunderstandings develop, creating problems. One common problem is the misunderstanding of quality standards or the reasoning behind them. Franchisees may not appreciate the methods used to maintain those standards or the inspection procedures used by the franchisors.

Franchisees may also develop a sense of independence and may not take advice of the franchisors. They may feel that they are better qualified than staff at the corporate offices. Contract language and further communications may be misinterpreted, resulting in a lack of cooperation by the franchisees. There may be personality differences between the regional staff of the franchisor and the franchisees.

Franchisees may be reluctant to disclose gross sales, on which royalty fees are dependent, or may not report accurately. Non-cooperation from franchisees may require adequate policing or may end up in long legal battles that may adversely affect an otherwise healthy business.

Pros and cons of franchising, from the franchisor's point of view, are outlined in Table 3.1. As in the case of franchisees, all advantages and disadvantages may not be present in a particular franchise system. Evidently, advantages far outweigh the disadvantages both from the franchisee's and franchisor's points of view.

Restaurant Study 4

Long John Silver's Restaurants, Inc.

Long John Silver's is the nation's leading quick-service seafood restaurant chain with more than 1,400 company-owned and franchise locations in 37 states, Singapore, and Thailand.

Long John Silver's employs approximately 26,000 team members systemwide. The corporate support center, where 165 persons are employed, is located in Kincaid Towers in downtown Lexington, Kentucky.

Long John Silver's is positioned for dramatic growth. Since 1990, Long John Silver's has crafted a corporate mission statement which effectively guides decision making at every level. This clearly focused corporate mission, along with a restructured, committed culture and an awareness of the marketplace, will enable the company to achieve its customer-driven goal.

Their promise is to provide each customer with great-tasting, healthful, reasonably priced fish, seafood, and chicken in a fast, friendly manner on every visit.

Their culture is to maintain a work environment that encourages team members to put forth their best efforts to serve customers. They will respect each team member as they work together to achieve excellence.

HISTORY

The first Long John Silver's Fish 'n' Chips opened in 1969, creating a totally new market for quick-service seafood. It was a pioneering effort to introduce seafood as a competitor to established American favorites such as hamburgers, pizza, and fried chicken.

Long John Silver's—the name was inspired by Robert Louis Stevenson's classic *Treasure Island*—found overwhelming acceptance for this new approach. The restaurant resembled a wharfside building, with brass lanterns, colorful oars, and signal flags decorating the interiors.

Today, exteriors have bolder colors, accent stripes, illuminated canopies on drive-thrus, and a new roof design—all of which give our restaurants a stronger retail identity.

As the early success of the concept became evident, Long John Silver's moved to expand the menu from its original batter-dipped fish and chicken peg legs to include a full line of seafood. Continuous innovation has brought customers sandwiches, salads, and a wide selection of side items.

Long John Silver's was a public company and traded in the over-the-counter market for 20 years as Jerrico Inc., a name derived from company founder Jerome Lederer. In September 1989, Jerrico and its sub-

(a)

(b)

An interior view of a Long John Silver's Restaurant (courtesy Long John Silver's, Inc.).

sidiaries were acquired for $620 million in a leveraged buy-out by senior management and a New York investment firm.

In 1990, the new privately held company divested itself of three restaurant concepts to devote its full resources to the operation of Long John Silver's. It is now known exclusively by its popular, branded name—Long John Silver's.

In fiscal 1997, the company had systemwide sales of $883.2 million—approximately 63 percent of the nation's $1.4 billion quick-service seafood business.

The company's marketing approach is designed to meet the needs of individual markets. A marketing services staff provides in-depth marketing research, sales analyses, product improvement, and new-product development.

Field marketing directors analyze market conditions and implement local marketing programs to make the most of business opportunities.

Long John Silver's is experiencing substantial annual franchise growth. Long John Silver's plans to add 50 franchise units per year to the chain within the next several years, focusing primarily on markets where Long John Silver's already has a presence.

An exterior view of the new Long John Silver's Restaurant in Lexington, KY (courtesy Long John Silver's, Inc.).

Currently, Long John Silver's has more than 90 participating franchise groups ranging in size from one to 100 units. An aggressive franchise plan includes business development in every region of the nation.

In addition to traditional freestanding restaurants, Long John Silver's offers exciting development opportunities in nontraditional locations, including convenience stores, freestanding food courts, and co-branded restaurants.

Restaurant Study 5

Godfather's Pizza

Godfather's Pizza, Inc., was established in 1973 and awarded its first franchise license agreement in 1974. Today Godfather's Pizza, Inc., has over 392 franchised restaurants and 180 company-owned restaurants operating in 40 states and Canada. The major goal of Godfather's Pizza, Inc., is to become the most profitable pizza company in the world—not the biggest chain. This will be achieved through becoming the world's most service-oriented pizza company. With this in mind, their main objective for growth is *quality growth* leading to *quality operations.* According to Godfather's Pizza, Inc., the crucial factor for achieving this quality growth is to find candidates who exhibit *characteristics for success* as franchisees. The most important characteristic is active "hands-on" involvement in the business. A second crucial factor to the success of Godfather's Pizza, Inc., is the continuing *cooperative relationship* it maintains with its franchisees. To achieve this, in principle and practice, Godfather's Pizza, Inc., places utmost importance on effective communication, which is the cornerstone of their management and franchise philosophy.

In 1985, Godfather's Pizza was in big trouble. The Omaha-based chain, which had recently been purchased by The Pillsbury Company, was plagued by steadily declining sales, unprofitable stores, rock-bottom morale at the restaurant level, and a stack of lawsuits filed against it by unhappy franchisees. The company was only partially joking when it said that it was considering installing a revolving door on the President's office, as the pizza chain had seen three Presidents come and go in less than a year. Herman Cain joined the beleaguered chain in 1986. Fresh from rejuvenating Burger King's Philadelphia region, the energetic and charismatic Cain set about turning the company around. He began closing unprofitable units, strengthening the company's advertising thrust, repairing franchisee relations, and quickly resolving legal conflicts. In fiscal 1987, Godfather's Pizza reversed a steady, three-year decline in restaurant sales, despite stiff competition. The secret to success, Cain revealed, was as simple as a renewed emphasis on what made Godfather's Pizza successful in the first place—consistent quality and service.

In 1988, Cain and Executive Vice President Ron Gartlan led a select

(a)

(b)

An exterior view of a Godfather's Pizza restaurant (courtesy Godfather's Pizza, Inc.).

group of Godfather's Pizza's senior management staff in purchasing the chain from Pillsbury and have continued on the same successful path they began in 1986; with Gartlan serving as President since 1995 and CEO since 1996.

Committed to carving a niche for itself as an industry leader, Godfather's Pizza embarked on an ambitious plan to retain its national reputation. New products, including such unique items as bacon-cheeseburger pizza, meatball pizza, and dessert pizza were successfully introduced. Also introduced were all-you-can-eat buffets and home delivery service. Godfather's Pizza now has more than 572 units in 38 states.

CHAPTER 4

Franchising Agreements and Legal Documents

Franchising is a legal contract between a franchisor and a franchisee. Therefore, all laws pertaining to contractual agreements apply to franchising. Additional laws pertain to franchising and deal with the franchisor-franchisee relationship; these exist at both the federal and the state level. The major trade regulation rule under the jurisdiction of the Federal Trade Commission (FTC) is referred to as the *Franchise Rule*, discussed in detail in the following paragraphs.

THE FRANCHISE RULE

Formally titled "Disclosure Requirements and Prohibitions Concerning Franchising and Business Opportunity Ventures," The FTC's Franchise Rule was adopted in response to widespread evidence of deceptive and unfair practices in connection with the sale of franchises. These practices may occur when prospective franchisees lack a ready means of obtaining essential and reliable information about their proposed business investment. This lack of information reduces the ability of prospective franchisees either to make an informed investment decision or otherwise to verify the representations of those selling the business. The Rule attempts to deal with these problems by requiring franchisors and franchise brokers to furnish prospective franchisees with information about the franchisor, the franchise business, and the terms of the franchise agreement. Although it requires disclosure of material facts, it does not regulate the substantive terms of the franchisor-franchisee relationship. The Franchise Rule became effective on October 21, 1979.

The provisions of the FTC's trade regulation rule are described in the interpretive guide entitled "Disclosure Requirements and Prohibitions Concerning Franchising and Business Opportunity Ventures." The three elements common to franchising covered by the rule are: (1) distribution of goods or services associated with the franchisor's trademark, (2) the franchisor's control over or assistance in the franchisee's method of operation, and (3) the franchisee's requirement to make a payment of $500 or more to the franchisor or a person affiliated with the franchisor at any time before or within six months of the business opening. These points are described below.

1. *Trademark.* This condition occurs when the franchisee is given the right to distribute goods and services that bear the franchisor's trademark, service mark, trade name, advertising, or other commercial symbol ("the mark"). The most common application of this condition occurs when either the goods or services being distributed by the franchisee are associated with the franchisor's mark, or when (a) the franchisee must conform to quality standards established by the franchisor with respect to the goods or services being distributed, and (b) the franchisee operates under a name that includes, in whole or in part, the fran-

chisor's mark. The FTC does not cover franchises in which no mark is involved.

2. *Significant Control or Assistance.* The term *significant* relates to the degree to which the franchisee is dependent on the franchisor's superior business expertise.

Among the significant types of control over the franchisee's method of operation are those involving (a) site approval for unestablished businesses, (b) site design or appearance requirements, (c) hours of operation, (d) production techniques, (e) accounting practices, (f) personnel policies and practices, (g) promotional campaigns requiring franchisee participation or financial contribution, (h) restrictions on customers, and (i) location or sales area restrictions.

Among the significant types of promises of assistance to the franchisee's method of operation are (a) sponsoring formal sales, repair, or business training programs, (b) establishing accounting systems, (c) furnishing management, marketing, or personnel advice, (d) selecting site locations, and (e) furnishing a detailed operating manual.

In order to be deemed significant, the controls or assistance must be related to the franchisee's entire method of operation—not its method of selling a specific product or products that represent a small part of the franchisee's business.

3. *Required Payments.* The franchisee must be required to pay to the franchisor (or an affiliate of the franchisor), as a condition of obtaining or commencing the franchise operation, a sum of at least $500 during a period from any time before to within six months after commencing operation of the franchisee's business. All types of payments are covered in this component of the requirement.

Franchisors falling under the abovementioned guidelines are required to furnish disclosure documents according to this rule. The Rule exempts (a) fractional franchises (a fractional franchise relationship results when an established distributor adds a franchised line to its existing line of goods), (b) leased department arrangements, (c) purely verbal agreements, and (d) minimal investments. It excludes (a) relationships between employer and employees, and among general business partners, (b) membership in retailer-owned cooperatives, (c) certification and testing services, and (d) single trademark licenses.

DISCLOSURE

Disclosure must be made by franchisors and franchise brokers. Disclosures must be made to prospective franchisees and, in some circumstances, to existing franchisees who extend or renew the term of their franchises. The disclosures must be made in a single document commonly known as the *disclosure document.* The document may not include information other than that required by the Rule or by state law not preempted by the Rule. However, the franchisor may furnish other information to the prospective franchisee that is not inconsistent with the material set forth in the disclosure document.

The Basic Disclosure Document

The basic disclosure document (often referred to as the *offering circular*) must be given to a prospective franchisee at the earlier of the first personal meeting or the time for making disclosures. The term *personal meeting* is defined as a face-to-face meeting between a prospective franchisee and a franchisor or franchise broker that is held for the purpose of discussing the sale, or possible sale, of a franchise. It does not include communication by telephone or mail. The term *time for making disclosures* is defined as ten business days prior to the earlier of (a) the execution of any franchise agreement, or (b) the payment by a prospective franchisee in consideration of the sale or proposed sale of a franchise.

A copy of the franchisor's standard franchise agreement and any related agreements, such as leases, purchase orders, etc., must be given to the prospective franchisee when the basic disclosure document is furnished. In addition, a copy of the completed franchise agreement intended to be executed by the parties must be given to the prospective franchisee at least five business days prior to the day that the agreements are to be executed.

To ensure that the disclosure documents contain accurate and timely information, the Rule requires that they be kept current. The basic disclosure document must be current as of the franchisor's most recent fiscal year. After the close of the fiscal year, the franchisor has ninety days in which to prepare a revised basic disclosure document. The document must also be updated at least quarterly whenever a "material" change occurs in the information it contains.

Uniform Franchise Offering Circular

Several states permit use of a disclosure format known as the Uniform Franchise Offering Circular (UFOC) to comply with their own state registration or disclosure requirements. The UFOC format was adopted by the Midwest Securities Commissioners Association on September 2, 1975. The FTC determined that, in the aggregate, the disclosures required by the UFOC format provide protection to the prospective franchisees that is equal to or greater than that provided by the Rule. Therefore, the FTC permits the use of the UFOC in lieu of its own disclosure requirements. For certain changes and limitations, the original guidelines should be consulted. Either the Rule or the UFOC disclosure format must be used in its entirety. Franchisors or franchise brokers may not pick and choose questions from each format. It should be understood that the FTC disclosure format is accepted in all fifty states, whereas the UFOC format is not. Also, the FTC requires disclosure of important facts but does not require registration, approval, or the filing of any documents with the FTC in connection with the sale of franchises.

Contents of the Disclosure Document

The basic disclosure document of the FTC consists of twenty categories of information, whereas the UFOC disclosure format contains twenty-three categories of information. Description of the categories of information required by the FTC is given below, along with a comparison to the UFOC format.

1. *Identifying Information about the Franchisor.* This includes:

- The official name and address and principal place of business of the franchisor and of the parent firm or holding company of the franchisor (if any)
- The name under which the franchisor is doing or intends to do business
- The trademarks, trade names, service marks, advertising, or other commercial symbols (the marks) that identify the goods, commodities, or services to be offered, sold, or distributed by the prospective franchises, or under which the prospective francIhisee will be operating

The UFOC requires the disclosure whether the franchisor is a corporation, a partnership, or any other type of business entity.

2. *Business Experience of Franchisor's Directors and Key Executive Officers.* Disclosure should include the name and relevant business experience of the franchisor's current directors as well as those executive officers who will have significant management responsibilities with respect to the marketing and servicing of franchises, such as the chief executive and operating officers, and financial, franchise marketing, franchise training, and franchise service officers.

Information for each listed person should include the person's current position and facts about his or her business experience during the preceding five fiscal years. This should include names of employers and positions or titles held. Other facts that would assist a prospective franchisee in assessing the person's business experience should be included. The UFOC also requires a disclosure of the names and business history of franchise brokers or subfranchisors who are affiliated with the franchisor and who will have management responsibility relating to the franchise.

3. *Business Experience of the Franchisor.* This requires disclosure of the business experience of the franchisor and the franchisor's parent firm (if any), including the length of time each (a) has conducted a business of the type to be operated by the franchisee, (b) has offered or sold a franchise for such business, (c) has conducted a business or offered or sold a franchise, and (d) has offered for sale or sold franchises in other lines of business, together with a description of such other lines of business.

4. *Litigation History.* This disclosure involves three types of franchisor litigation: criminal, civil, and administrative. For each past criminal, civil, or administrative proceeding disclosed, (a) the identity of the court or agency involved, (b) the date of conviction, judgment, decision, order, or ruling, (c) the amount of award or judgment, and (d) the terms of any settlement order or ruling should be included. This pertains to the franchisor and any person identified in items 2 and 3 above. Both the FTC Rule and the UFOC require litigation history disclosure. The UFOC requires disclosure for related subfranchisors and franchise brokers as well. The FTC Rule requires a disclosure of litigation over the past seven fiscal years, whereas the UFOC requires litigation history for the past ten fiscal years. Also, the FTC Rule requires disclosure of litigation in the United States only, while the UFOC also includes litigation in Canada.

5. *Bankruptcy History.* Disclosure involves bankruptcy history during the previous seven fiscal years (UFOC requires disclosure of bankruptcy history for the past ten fiscal years) of the franchisor, its parent or holding company, and those of its current directors and executive officers for whom disclosure is required. The franchisor should disclose, with respect to each bankruptcy proceeding requiring disclosure: (a) the name of the person(s) or business entity who has filed in bankruptcy, been adjudged bankrupt, or been re-

organized due to insolvency (if other than the franchisor, the identity of such persons should be disclosed, e.g., the chief executive officer), (b) the court in which the proceeding was held, including the case or docket number, and the nature of the proceeding, such as bankruptcy, reorganization, etc., (c) the date of adjudication of bankruptcy or confirmation of a plan for reorganization and the date of discharge, and (d) any material facts. This section does not require disclosure where the franchisor or an appropriate officer merely files a claim in a bankruptcy proceeding.

6. *Description of Franchise.* Disclosure should provide a factual description of the franchise being offered for sale. The term *sold* encompasses purchasing, leasing, licensing, and other methods of acquisition. Included within this disclosure should be (a) a general description of the business to be conducted by the franchisee, (b) a detailed discussion of the business format and/or product line that the franchisee is purchasing, including goods or services to be sold by the franchisee, and (c) a description of the market for the goods and/or services to be sold by the franchisee (e.g., whether the goods will be marketed to a specific segment of the community such as student, elderly, upper income consumers).

7. *Initial Funds Required to Be Paid by a Franchisee.* The franchisor must disclose the nature, amount, payee, and due date of all monies that the franchisee must pay in order to obtain or commence the franchise operation, in those circumstances when such payment is made either to (a) the franchisor, (b) an affiliate of the franchisor, or when (c) the franchisor or affiliate collects the payment on behalf of a third party. Such payments include, but are not limited to, the initial franchise fees, deposits, down payments, prepaid rent, and equipment and inventory purchases. If exact amounts for any one or more categories vary, then a reasonable range of anticipated payments for each of such categories may be substituted. The franchisor need not disclose required payments that the franchisee must make to obtain or commence the franchise business (such as payments for the purchase of signs) when the franchisee has the option to make such payments to unaffiliated third parties.

The statement must indicate, for each payment, whether all or any part is refundable, and, if so, under what conditions. Although the Rule neither requires nor prohibits refund promises, it does require that refunds must be made when promised. Failure to do so constitutes a violation of the Rule.

Franchisors also must disclose any nonrecurring commitments of funds by franchisees to the franchisor or affiliated persons for securing the franchise if the commitment is made when the franchise relationship is commenced, even though payment is not required until a later date, such as a twelve-month deferred payment of a part of the franchise fee.

It is recommended, though not required by the Rule, that franchisors disclose the nature and approximate amount of other payments that a franchisee must make to obtain or commence business. The UFOC also requires disclosure of suggested working capital requirements. The UFOC also requires the franchisor to disclose whether identical franchise fees or initial payments are charged for each franchise. In cases where fees are not identical, a statement of the formula or method for determining the amount of the fee must be disclosed. It also requires the franchisor to disclose how it will use payments it receives.

8. *Recurring Funds Required to Be paid by a Franchisee.* The franchisor must disclose the nature, amount, and payee of all payments that a franchisee must make on a recurring basis in carrying on the franchise business in those circumstances when those payments are made to (a) the franchisor, (b) an affiliate of the franchisor, or when (c) the franchisor or its affiliate collects the payment on behalf of a third party. Such payments include, but are not limited to, royalty, lease, advertising, training, sign rental fees, and equipment purchases. Two categories of recurring payments should be listed: those payable on a regular periodic basis, such as royalties, advertising, and inventory purchases (in those circumstances where there are minimum purchase requirements), and those infrequent anticipated expenses of a major nature, such as the replacement cost of expensive equipment.

The amount of the payments should be expressed in an estimated dollar amount. If the amount of payment is dependent on a variable, such as sales volume, then these amounts may be expressed as a percentage of such variable. Where no accurate dollar amount is available, an estimated payment range may be used. Infrequent anticipated major expenses may be expressed either in their present or estimated future cost. The UFOC format requires additional disclosure of whether any recurring or isolated fees are refundable.

9. *Affiliated Persons the Franchisee is Required or Advised to Do Business With by the Franchisor.* This disclosure requires a list of persons, who are either the franchisor or any of its affiliates, with whom the franchisee is required or advised to do business. The franchisor must list those suppliers who it requires or advises the franchisee to use, regardless of whether the supplier is the sole approved supplier or one of several approved suppliers, whenever such supplier is either the franchisor or an affiliate of the franchisor. The supplier should be listed even if its use is recommended rather than required. A brief description of the goods or services supplied by any listed supplier also must be disclosed.

10. *Obligations to Purchase.* This disclosure requires a description of specified items related to the establishment or operation of the franchise business that the franchisor requires the franchisee to purchase, lease, or rent. Such items include real estate, services, supplies, products, inventories, signs, fixtures, and equipment. If any listed items must conform to franchisor-imposed specifications, such as brand names or product standards, the existence of such specifications must be disclosed. If such specifications make the item substantially more expensive or difficult to obtain, this should be mentioned. The franchisor should indicate the manner in which the franchisor issues and changes specifications as well as the business justification for such specification(s). The UFOC format also requires disclosure of the amount of the purchases the franchisor requires the franchisee to make.

The list of required purchases and required suppliers may be contained in a document separate from the basic disclosure document, if the separate document is delivered to the prospective franchisee along with the basic disclosure document, and if the basic disclosure document notes the existence of such other document.

11. *Revenue Received by the Franchisor in Consideration of Purchases by a Franchisee.* This involves a description of consideration paid (such as royalties or commissions) by third parties to the franchisor or any of its affiliates as a result of a franchisee's purchase from such third parties. Disclosure is limited to situations in which (a) the supplier (or group of suppliers) is a required or advised source of franchisee purchases, and (b) the rebate is received by the franchisor as a result of such purchases by the franchisee. The term *rebate* refers to any revenue or other monetary or nonmonetary consideration, including but not limited to cash payments in the form of kickbacks or commissions.

The franchisor must disclose both the basis for calculating rebates and the amount received by them or their affiliates.

12. *Financing Arrangements.* This disclosure includes a description of any franchisor assistance in financing the purchase of a franchise. The UFOC format requires more detailed disclosure than does the FTC Rule. The UFOC format also requires examples of the legal documents in which the financing arrangements are set forth.

Disclosure description should include the material terms of financing arrangements offered to the franchisee where such financing is to be made directly by the franchisor or any affiliated person, or indirectly through third parties who lend money to franchisees as a result of an arrangement with or through the intercession of the franchisor. Materials terms include items such as the name and address of the lender, the amount to be financed, the terms and annual percentage interest rates, repayment rights, and provisions in the event of default. The franchisor must disclose any rebate received by it, or an affiliate, from a third person arising out of or in consideration of a franchisee's financing arrangement, such as a finder's fee.

Neither open account financing payable within ninety days nor franchise fees payable, without interest, over a period of time need be disclosed under this section.

13. *Restriction of Sales.* This disclosure requires a description of whether the franchisee is: (a) limited in the type of products or services it may sell, (b) limited in the customers to whom it may sell, (c) limited in the geographic area in which it may sell, or (d) granted territorial protection by the franchisor.

Any of the foregoing limitations or grants may result from specific terms of the franchise agreement or by written or verbal understanding.

The disclosure must describe the specific limitation(s) involved and the franchisor's reason(s) for imposing such limitation(s). The description of any geographic limitation on a franchisee should include the typical boundaries of such area. If the franchisee is limited to selling goods or services from a particular location, this fact should also be disclosed. The UFOC format requires a statement of whether or not sales goals must be achieved to maintain territorial limitations.

14. *Personal Participation Required of the Franchisee in the Operation of the Franchise.* A statement should be included in the document related to the extent to which the franchisor requires the franchisee (or, if the franchisee is a corporation, any person affiliated with the franchisee) to participate personally in the direct operation of the franchise. A brief description of the types of activities that constitute such participation should be included. In the case of a corporation, the statement should indicate whether any specific director or employee thereof must personally participate in the direct operation of the franchise business.

15. *Termination, Cancellation, and Renewal of the Franchise.* This requires a statement disclosing, with respect to the franchise agreement and any related agreements:

   a. The term (if any) of such agreement, and whether such term is or may be affected by any agreement (including leases or subleases)

b. The conditions under which the franchises may renew or extend
c. The conditions under which the franchisor may refuse to renew or extend
d. The conditions under which the franchisee may terminate
e. The conditions under which the franchisor may terminate
f. The obligations of the franchisee after termination of the franchise by the franchisor, and the obligations of the franchisee after termination of the franchise by the franchisee and after the expiration of the franchise
g. The franchisee's interest upon termination of the franchise, or upon refusal to renew or extend the franchise, whether by the franchisor or by the franchisee
h. The conditions under which the franchisor may repurchase, whether by right of first refusal or at the option of the franchisor
i. The conditions under which the franchisee may sell or assign all or any interest in the ownership of the franchise, or of the assets of the franchise business
j. The conditions under which the franchisor may sell or assign, in whole or in part, its interest under such agreements
k. The conditions under which the franchisor may modify
l. The conditions under which the franchisee may modify
m. The rights of the franchisee's heirs or personal representative upon the death or incapacity of the franchisee
n. The provisions of any covenant not to compete

16. *Statistical Information Concerning the Number of Franchises and Company-Owned Outlets.* This disclosure requires statements as to the total number of operating franchises and company-owned outlets of the franchisor, as well as the number of franchises which the franchisor has terminated, failed to renew, or reacquired during the preceding fiscal year. It also requires disclosure of the number of franchises voluntarily terminated or not renewed by the franchisee. The franchisor may comply by (a) listing the ten fran-

chised outlets nearest the prospective franchisee's intended location (or all franchise units, if fewer than ten), (b) listing all franchisees, or (c) listing all franchisees located in either the state where the prospective franchisee will locate its business or where the prospective franchisee lives. General reasons such as "failure to comply with quality control standards" or "failure to make sufficient sales" should be provided wherever applicable. The franchisor is not required to disclose either the name or any other identifying information about any terminated franchisee. The UFOC format requires disclosure of franchises that have been sold but are not yet in operation and an estimate of the number of franchises to be sold during the coming year. While the FTC rule requires information pertaining to preceding fiscal year, the UFOC format requires that for the preceding three fiscal years.

17. *Site Selection.* The disclosure required by this section concerns the selection or approval of a site for the proposed franchise outlet and the time frames for such activity, based on the franchisor's experience in the preceding fiscal year.

18. *Training Programs.* If the franchisor offers, an initial training program or informs the prospective franchisee that it intends to provide such person with initial training, a statement disclosing (a) the type and nature of the initial training, (b) the minimum amount, if any, of training that will be provided, and (c) the cost, if any, to be borne by the franchisee for such training, should be included. The type and nature of the training should include a description of the general contents of the initial training program and all elements of such training. The required disclosure is limited to the training the franchisor offers at the beginning of the franchise relationship—that is, from the period after the execution of the franchise agreement through shortly after the actual commencement of the franchise business. The franchisor, at its option, may describe any additional training available to franchisees during the term of the franchise relationship. The UFOC requires detailed disclosure of training aspects such as the duration, content, cost of training programs, and training experience of the instructors.

19. *Public Figure Involvement in the Franchise.* If the name of a public figure is used in connection with a recommendation to purchase a franchise, or as a part of the name of the franchise operation, or if the public figure is stated to be involved with the management of the franchisor, a statement disclosing (a) the nature and extent of the public figure's involvement and obligations to the franchisor, including but not limited to the promotional assistance the public figure will provide to the franchisor and to the franchisee, (b) the total investment of the public figure in the franchise operation, and (c) the amount of any fee or fees the franchisee will be obligated to pay for such involvement or assessment provided by the public figure.

The term *public figure* refers to a person whose identity would be known to a substantial portion of the public either nationally or within the geographic area in which the franchise is sold, such as a person who has achieved prominence in sports, entertainment, or public affairs. The term does not include nonliving or fictionalized characters.

20. *Financial Information Concerning the Franchisor.* Required disclosure includes a balance sheet (statement of financial position) for the franchisor's most recent fiscal year, an income statement (statement of results of operations) for the most recent three fiscal years, and a statement of changes in financial position for the most recent three fiscal years. Financial statements prepared and filed with the Securities and Exchange Commission (SEC) in accordance with SEC Regulation S-X and the SEC's Accounting Series Releases may be used. Although audited financial statements are required, under certain conditions unaudited financial statements may be used. Updated information should be provided to prospective franchisees when material changes occur in the information contained in the financial statements.

As evident, disclosure document requirements of the FTC Rule and the UFOC are similar. The UFOC format requires detailed disclosure about patents and copyrights that are part of the franchise system. A sample copy of the franchising agreement is required to be *attached* to the offering circular under the UFOC guidelines, whereas the FTC Rule requires that proposed agreements *accompany* the offering circular. All contractual agreements under both formats are required to be delivered to the prospective franchisee at least five business days prior to execution. Under the UFOC guidelines, the last page of each offering circular is a detachable acknowledgment of receipt. The prospective franchisee returns it to the franchisor after signing as an acknowledgment of the date of receipt of the offering circular. An example of such an acknowledgment is shown in Figure 4.1.

All of the foregoing information required by the FTC Rule shall be contained in a single disclosure statement or prospectus, which shall not contain any materials or information other than that required. This does not preclude franchisors or franchise brokers from giving other nondeceptive information orally, visually, or in separate literature so long as such information is not contradictory to the information in the disclosure statement required by the FTC Rule. This disclosure statement shall carry a

---

ITEM XXIII

**ACKNOWLEDGMENT OF RECEIPT
BY PROSPECTIVE LICENSEE**

The undersigned, personally or as an officer or a partner of the proposed Licensee, does hereby acknowledge receipt of the "Uniform License Offering Circular" dated *May 1, 1989* (to which this receipt is attached), of Hardee's Food Systems, Inc. required by this state, the Federal Trade Commission, or both, and of the earnings claims, financial statements, list of licensees, and Hardee's Food Systems, Inc. License Agreement (and addendum) which are included therein.

_____   _____
Date of Receipt   Individually or as an Officer
                                                                        of_____
                                                                        (a_____Corporation)
                                                                        (a_____Partnership)

FIGURE 4.1 An example of the acknowledgment required by FTC Rule (courtesy Hardee's Food System).

cover sheet distinctively and conspicuously showing the name of the franchisor, the date of issuance of the disclosure statement, and the notice, as shown in Figure 4.2.

All information contained in the disclosure statement shall be current as of the close of the franchisor's most recent fiscal year. After the close of each fiscal year, the franchisor is given a period not exceeding ninety days to prepare a revised disclosure statement. A table of contents is included within the disclosure statement. The disclosure statement includes a comment that either positively or negatively responds to each disclosure item required to be in the disclosure statement.

The disclosure document must be given to a prospective franchisee at the earlier of either (1) the prospective franchisee's first personal meeting with the franchisor, or (2) ten days prior to the execution of a contract or payment of money relating to the franchise relationship. In addition to the document, the franchisee

---

HARDEE'S FOOD SYSTEMS, INC.

INFORMATION FOR PROSPECTIVE FRANCHISEES
REQUIRED BY FEDERAL TRADE COMMISSION

*****

To protect you, we've required your franchisor to give you this information. *We haven't checked it, and don't know if it's correct.* It should help you make up your mind. Study it carefully. While it includes some information about your contract, don't rely on it alone to understand your contract. Read all of your contract carefully. Buying a franchise is a complicated investment. Take your time to decide. If possible, show your contract and this information to an advisor, like a lawyer or an accountant. If you find anything you think may be wrong or anything important that's been left out, you should let us know about it. It may be against the law.

There may also be laws on franchising in your state. Ask your state agencies about them.

<u>Federal Trade Commission</u>
<u>Washington, D.C.</u> 20580

Date of Issuance: _____<u>May 1</u>_____, 1989

FIGURE 4.2 An example of the front page on disclosure document required by FTC Rule (courtesy Hardee's Food System).

must receive a copy of all agreements that he or she will be asked to sign.

Earnings Claim

The Rule prohibits earnings representations about the actual or potential sales, income, or profits of existing or prospective franchisees unless:

1. Reasonable proof exists to support the accuracy of the claim.
2. The franchisor has in its possession, at the time the claim is made, information sufficient to substantiate the accuracy of the claim.
3. The claim is geographically relevant to the prospective franchisee's proposed location.
4. An earnings claim disclosure document is given to the prospective franchisee at the same time that the other disclosures are given. The earnings claim document must contain six items:
   a. Cover sheet, as specified in the Rule
   b. The earnings claim
   c. A statement of the bases and assumptions on which the earnings claim is made
   d. Information concerning the number and percentage of outlets that have earned at least the amount set forth in the claim, or a statement of lack of experience, as well as the beginning and ending dates of the time period covered by the claim
   e. A mandatory caution statement, whose text is set forth in the Rule, concerning the likelihood of duplicating the earnings claim
   f. A statement that information sufficient to substantiate the accuracy of the claim is available for inspection by the franchisee

Prospective franchisees must be notified of any material changes in the information contained in the earnings claim document prior to becoming a franchisee.

VIOLATION OF THE FRANCHISE RULE

It is an unfair or deceptive act or practice within the meaning of the FTC Act for any franchisor or franchise broker:

1. To fail to furnish prospective franchisees, within the time frames established by the Rule, with a disclosure document containing information on twenty different subjects relating to the franchisor, the franchise business, and the terms of the franchise agreement
2. To make any representations about the actual or potential sales, income, or profits of existing or prospective franchises except in the manner set forth in the Rule
3. To make any claim or representation (such as in advertising or oral statements by salespersons) that is inconsistent with the information required to be disclosed by the Rule
4. To fail to furnish prospective franchisees, within the time frames established by the Rule, with copies of the franchisor's standard forms of franchise agreements and copies of the final agreements to be signed by the parties
5. To fail to return to prospective franchisees any funds or deposits (such as down payments) identified as refundable in the disclosure document.

Violators are subject to civil penalty actions brought by the FTC of up to $10,000 per violation.

STATE FRANCHISE LAWS

The Federal Trade Commission's goals are to create a minimum federal standard of disclosure applicable to all franchisor offerings, and to permit states to provide additional protection as they see fit. Thus, while the FTC trade regulation rules have the force and effect of federal law and, like other federal substantive regulations, preempt state and local laws to the extent that these laws conflict, the FTC has determined that the Rule will not preempt state or local laws and regulations that either are consistent with the Rule or, even if inconsistent, that would provide protection to prospective franchisees equal to or greater than that imposed by the Rule. Examples of state laws or regulations that would not be preempted by the Rule include state provisions requiring the registration of franchisors and franchise salesmen, state requirements for escrow or bonding arrangements, and state-required disclosure obligations exceeding the disclosure obligations set forth in the Rule. Moreover, the Rule does not affect state laws or regulations that regulate the franchisor-franchisee relationship, such as termination practices, contract provisions, and financing arrangements. Several states have their own franchise disclosure laws.

THE FRANCHISE AGREEMENT

The Franchise Agreement is the most important document for both franchisors and franchisees. Legal counsel is necessary when entering into any agreement. The following items constitute the franchise agreement of a typical franchise, although this list is not comprehensive and agreements vary to a considerable extent:

1. *Introduction.* In this item are listed the name of the franchise, the date of agreement, the nature of the franchise system, the identity of the names, and symbols associated with the franchise. Also included in this section are the identity of the parties to the agreement and the location of the business.
2. *Duration of the Agreement.* The length of the initial term of the agreement and the starting date should be included

in this section. Also included are clauses pertaining to the renewal terms and related conditions.
3. *Fees and Other Payments.* The initial franchise fee and other payments (such as fees for promoting the opening of the restaurant) should be included under this item. Also recurring fees (such as royalty fees, advertising and promotion fees, and service fees) to be paid (normally as the percentage of the gross sales) should be included. The types and terms of payments and whether these payments are refundable or nonrefundable should be specified. Fluctuations in the amount of fees to be paid or any adjustments should be indicated. What is meant by "gross sales" or "sales revenue" should be defined clearly in order to avoid later confusion.
4. *Responsibilities of the Franchisor.* This section should include types of services to be provided by the franchisor. These may include items such as site selection counseling and assistance, building plans and specifications, plans and specifications of improvements, operational training, pre-opening and opening assistance, advertising and marketing assistance, continuing communications and counseling, accounting procedures, and provision of a complete operating manual. Also specified should be the steps taken by the franchisor to maintain high and uniform standards of quality, cleanliness, appearance, and service. Details of inspection procedures and periodic evaluations of the products and services provided should be included in the documents submitted.
5. *Responsibilities of the Franchisee.* Expectations from the franchisee should be specified in this section. Items to be included are improvements on the property, supervision by the franchisor, construction of the facilities, need for required permits and certificates, and site layout and plans for construction or remodeling. Responsibilities also include attendance at and successful completion of the franchisor's training program. The use of the franchised premises solely

for the operation as required by the franchisor is specified here, as are details of the maintenance of the facility and sanitation requirements. Clearly set out are agreements pertaining to the operation of the restaurant in conformity with uniform methods, standards, and specifications set by the franchisors. Spelled out are guidelines for the purchase of food and supplies, inventory maintenance, and specifications pertaining to the purchase of those items. Also, the right to enter the premises by the franchisor or the franchisor's inspectors is described in detail in this section. This involves the quality control measures to both products and services offered by the franchisee.

6. *Proprietary Marks.* Trademarks, service marks, and other symbols play an extremely significant role in franchising. The agreement clarifies the franchisor's exclusive rights, title ownership, and proper display and use of the proprietary marks by the franchisee. In other words, this agreement confirms the proprietary rights of the franchisors and deals with an agreement on the use of proprietary marks solely in the manner prescribed by the franchisor.

7. *Operating Procedures and Confidentiality.* Confidentiality and adherence to the operating procedures are spelled out under this item. Franchising involves trade secrets that are the backbone for success of many franchises and therefore require the utmost security. Operating manuals are designed by franchisors to outline all details pertaining to the operations and quality of the products and services provided by the franchisees. The confidentiality and use of the operational manual by the franchisees are discussed under this item. Also, the rights of the franchisor to that manual and for its modification, if and when needed, are clarified.

8. *Advertising and Promotion.* Standardization of advertising and promotion is important in franchising. The maintenance of integrity and quality of promotional efforts is essential for success. Expenses involved and procedures for local, re-

gional, national, and international advertising should be recorded in this part of the agreement. Some of these procedures are also outlined in the operating manuals.
9. *Financial Records.* This section deals with the maintaining and preserving of records, books, accounts, and tax returns, which may be audited by the franchisor. Necessary forms and accounting procedures are outlined by the franchisors. This constitutes an important part of the agreement and both franchisor and franchisee should give careful consideration to the clause(s) under this item.
10. *Training.* The franchise agreement should include details of the training program as to the type of training, location, duration, and expenses to be paid by the franchisee. Also included should be the information on initial and continuing training programs and the person(s) required to be trained. Training responsibilities of the franchisor and the franchisee are discussed in an earlier section.
11. *Insurance.* All aspects related to the types of insurance and insurance cost should be detailed in this section. Many franchisors require that the franchisee carry an insurance policy prior to the opening of a restaurant, protecting both the franchisor and the franchisee and their officers, partners, and employees against any loss, liability, fire, lightning, earthquakes, personal injury, death, theft, vandalism, malicious mischief, property damage, and other possible calamities. The type of insurance policies needed and the agents from which these insurance policies can be purchased are described in this section. Details of modification in the policies and time frame for renewal should be specified.
12. *Products and Services to Be Purchased by the Franchisee.* If the franchisor requires the purchase of products and services not described earlier in the document, this should be included in this section. The type of products to be purchased and specifications required should be described in detail.

13. *Transferability of the Franchising Agreement.* Details of the conditions under which the agreement can be transferred are described in this clause. Franchising agreements are given to individual franchisees and are based on the skills and capabilities of the person or persons involved. Thus, permissions required or conditions to be fulfilled if the franchisee would like to sell, assign, transfer, convey, give away, pledge, mortgage, or otherwise encumber any interest, should be clearly stated. Provisions in special circumstances such as death, divorce, bankruptcy, and permanent or temporary disablement should be included. Procedures required for approval and training of the transferee should be specified. A new franchise agreement may be necessary in some instances and the provisions and financial requirements may have to be changed. State and local laws and regulations should be taken into account when considering this section of the agreement.
14. *Termination.* The rights to terminate this agreement by either party should be stated under this section. Termination is a part of the agreement that is frequently contested by both franchisors and franchisees and so should be carefully written and understood by both parties. Notices required and their contents should be specified. Types of default, steps to be taken to correct them, and periods given for rectification should be included. Conditions that would lead to termination should be stated in general as well as specific terms. The most common reasons for termination are:
    a. Failure or refusal or negligence in paying any monies owed to the franchisors
    b. Failure to follow terms agreed upon in the agreement
    c. Failure to maintain the quality control standards set by the franchisor
    d. Failure to complete training program specified by the franchisor

e. Continuous violation in connection with the operation of the restaurant of any law, ordinance, rule, or regulation of any governmental agency
   f. Improper use of the proprietary marks of the franchisor
   g. Certain conditions of bankruptcy and insolvency
   h. Franchisor's decision to withdraw from the marketing area

   The above-mentioned reasons are given as examples and situations may differ in individual cases. Reasons and consequences of termination should be described in this section of the agreement. Obligations of the franchisor and the franchisee on termination should be clearly stated.
15. *Covenants Not to Compete.* Franchisors may require that the franchisee devote full time, energy, and efforts to the management and operation of the franchisor's business at the specified location. There may be restrictions to getting into competing business.
16. *Renewals.* Absolute or conditional rights to renew may be included in the franchise agreement. Periods of time and conditions are to be specified.
17. *Miscellaneous Provisions.* Several other provisions could be included in the agreement. Some of them are:
    a. Arbitration
    b. Rights of first refusal
    c. Applicable law and its jurisdiction
    d. Severability and construction
    e. Notices
    f. Waivers and nonwaivers

Information included in this section of the chapter is intended to highlight the contents of the typical franchise agreement. Actual agreements may vary in content.

Restaurant Study 6

Arby's

THE HISTORY OF ARBY'S

Arby's was founded in 1964, but the concept for a roast beef sandwich franchise was rooted many years earlier. Forrest Raffel, a graduate of Cornell University School of Hotel and Restaurant Administration and his younger brother Leroy, a graduate of Wharton School of Finance, University of Pennsylvania, bought an uncle's restaurant equipment business in the 1950s. Raffel Brothers, Inc. became a fitting introduction to the foodservice industry for the brothers.

The small company rapidly grew to become one of the country's leading foodservice consulting firms. Raffel Brothers, Inc. designed and installed hundreds of foodservice facilities including the flight kitchens at Greater Pittsburgh International Airport, interiors of six Ohio Turnpike restaurants, and foodservice facilities for the Hospitality Inn motel chain of Standard Oil of Ohio.

But the Raffel brothers quickly sensed the potential of fast food and decided to develop a franchise operation based on something other than hamburgers. Leroy remembers, "We were totally confident, while everyone else thought we were out of our minds."

A late-night excursion to a small Boston sandwich shop one rainy Halloween was the inspiration for the Raffels, who joined a damp, but determined huddle of patrons to await the main attraction—a 79-cent roast beef sandwich.

The idea was born. The only kink in the chain's orderly development came with choosing a name for it. The partners wanted to use the name "Big Tex," but were unsuccessful in negotiating with the Akron businessman who was already using the name. So, in the words of Forrest, "We came up with Arby's, which stands for R.B., the initials of Raffel Brothers, although I guess customers might think the initials stand for roast beef."

The first Arby's unit opened in Boardman, Ohio, on July 23, 1964, serving only roast beef sandwiches, potato chips, and a beverage, and was the only store for a year as the brothers refined the operation. One year later, the first Arby's licensee opened a store in Akron, Ohio. Today, more than 500 licensees operate more than 3,000 restaurants worldwide.

Factors Responsible for the Success of Arby's, Inc.

1. Strong brand recognition with loyal consumers.
2. High-quality products and our commitment to delivering Cut-Above food.
3. The entrepreneurial spirit of the franchisees through development of new products and processes that improve the operations of the restaurants.
4. An improved relationship with the franchise community.
5. No direct ownership of the stores by TRG allows the franchisor organization to focus on the success of the franchisees.

Negative Factors and the Steps Taken to Rectify Them

- Lack of store distribution. Many parts of the country are without an Arby's.
  Resolution: To work with existing and new franchisees to build more stores. Current commitments total almost 1,000 new stores in the foreseeable future. Facilitating growth through work on economics of the building and alignment with several financial organizations to facilitate access to capital for the operators.
- Inconsistent leadership from the franchisor presented itself during two periods—the early 1970s and the early 1990s.
  Resolution: Solid senior management has been in place for 18 months and is committed to providing value-added services designed to help the franchisees maximize potential success. Strong relationships with the franchisee community have been given a higher priority status under current management.
- Poor reinvestment activity by franchisees and company in remodeling had resulted in hundreds of outdated facilities, which are still in operation.
  Resolution: Franchisor now offers two Remodeling Incentive Programs to spur facility upgrades. Franchisor is also aggressively enforcing operational and image covenants contained in the License Agreements.
- Inadequate media weights and ad spending, and inconsistent marketing efforts resulted in lower reach and effectiveness ratings.
  Resolution: AFA (System's marketing organization) is developing much stronger creative, using celebrity voice-overs in "Food Is the Hero"—themed copy. The marketing messages have been very well received by consumers and franchisees and are helping to drive sales volumes throughout 1998.

116 Restaurant Franchising

(a)

(b)

An exterior view of an Arby's restaurant (courtesy Arby's Inc.).

Philosophy of Operation

As an organization, TRG's mission statement is to become a world-class franchisor of high-quality restaurant brands. As part of that mission, their goal is to support their franchisees and to create a professional environment that remains focused on that goal. The success is measured by the success of their franchisees.

TRG is also committed to its employees and has developed a set of core values, which include Integrity, Flexibility, Accountability, Leading by Example, Eliminating Barriers, Communicating, Frugality, Team-Oriented, Flat, and Fun. As part of that commitment, TRG has made significant progress in terms of providing an above-industry-average compensation and bonus structure, expanded health benefits, and company recognition programs. Senior management has also commissioned the creation of a Presidential Advisory Council, which includes representatives from several departments of TRG, in an effort to address and solve internal problems.

Statement Related to Franchising

Over the 34 years in which the Arby's system has provided their customers with great-tasting products, Arby's, Inc. has experimented with a wide variety of franchising initiatives. Through those experiences they have determined that the plan of responsible system growth can best be delivered by providing our existing franchisees with unprecedented levels of service and support which will encourage them to grow their businesses.

Further, they believe that by attracting a small number of new franchisees each year, they will keep adding new members to the Arby's family who add fresh ideas and creative approaches to how to conduct the business of Arby's in the future.

Finally, they no longer sell franchises to anyone who can meet a certain financial worth ratio. Instead, they are in the business of granting and awarding development rights to qualified organizations that are interested and willing to grow their business in accordance with their License Agreements. This will ultimately bode extremely well for them, for TRG, and for their existing Franchisees.

CHAPTER 5

Franchisee/ Franchisor/Franchise Selection

Franchisee, franchisor, and *franchise* represent three sides of a triangle. Because the relationship in franchising is symbiotic, a preselection screening for each of the constituents is important. A franchisee has to be carefully selected by a franchisor to ensure mutual cooperation for the success of the franchise. The franchisee has to determine whether or not the franchise system offered is suitable for investment and if the franchisor will meet his or her expectations. Both franchisor and franchisee have to consider the suitability of the entire franchise system in which they are interested. All of these factors are considered in detail in this chapter.

SELECTION OF THE FRANCHISEE BY THE FRANCHISOR

There is no typical profile of a successful franchisee. If such a profile existed, all franchises would be successful. Successful franchisees are of all ages and both sexes. Franchisees are primarily entrepreneurs who have a desire to go into business for themselves. This does not necessarily mean that they are adaptable to any type of business or any type of franchise. Entrepreneurs may have enough money to invest but they may not be suitable for a particular franchise. A successful restaurant franchise may get many times more applications from prospective franchisees than there are franchises available. Therefore, it is essential to establish criteria for franchisee selection. Each franchisor has its own method of franchisee selection. Typical qualifications desirable by restaurant franchisors are discussed below.

Franchisee Qualifications

It is unrealistic to assume that all of the qualifications discussed below must be present in a prospective franchisee. However, these attributes help in setting and following a procedure for the selection of franchisees.

1. *Overall Business Experience.* Previous business experience of the franchisee is considered a primary factor in selection by many franchisors. This experience need not necessarily be in the restaurant industry. In fact, many successful restaurant franchisees were involved earlier in other types of businesses. Because restaurant business involves people, a strong business background with special emphasis on people-handling skills and management is desirable. Experience is preferably in human resources management—recruiting, training, supervising, and communicating. Highly desirable is previous experience in multi-unit chain operations. Familiarity with day-to-day retail business operations, preferably in restaurants, is an asset.
2. *Financial Qualifications.* Although business experience is the foremost qualification, it is not the only one desired in a franchisee. Franchising requires substantial investment

from franchisees; therefore, their financial qualifications are carefully assessed. The prospect must be financially able to provide the initial cash investment, with additional resources available thereafter, particularly for a possible financial emergency. Franchisors normally assess the net worth of an individual, excluding some personal possessions, such as home and car. Many franchisors provide a range of minimum investment, which may vary from location to location. Some franchisors offer prospective franchisees the options of leasing an existing restaurant, leasing a new restaurant, and purchasing a new or existing restaurant, at different investment costs.

Although it is not expected that all finances required should be paid by franchisees, the financial assessment also indicates the borrowing capacity of a prospective franchisee. Anywhere from 20 to 40 percent of the total cost is typically required to be funded from nonborrowed personal resources. The financial capability for further development is also assessed, particularly to see that the capital is not spread too thin in the franchisee's other undertakings. To determine financial status, confidential statements are an integral part of the franchise application. They include, primarily, a listing of personal assets, liabilities, and net worth. Additional information desired includes cash on hand and in the bank; securities; bonds and debentures; notes, accounts, and mortgages receivable; loans, notes, and accounts payable; stocks and bonds; cash value of life insurance; real estate owned; cash value of equipment and furniture; taxes due; business interests; other assets; and other debts and liabilities. Details of this collected information are shown in Appendices B and C.

3. *Proven Track Record.* The track record of previous undertakings and business ventures provides a valuable assessment of the franchisee's capacity to succeed. Success in one venture does not guarantee success in others, but it does provide an idea of the entrepreneurial nature of the franchisee.

Franchisees with a proven record in business have been shown to be better franchisees than those new in business.
4. *An Entrepreneurial Spirit and a Strong Desire to Succeed.* Franchising demands an entrepreneurial spirit and a strong desire to succeed on the part of all parties involved. It is not only the efficient running of the restaurants but the motivation to succeed that makes a franchisee successful. Franchisees should possess general business knowledge and be familiar with the operational aspects of a business. For example, it is important for a successful franchisee to be able to deal with the local community, bargain for choice locations, be an effective communicator, be aware of the local zoning laws and building codes, and, as well, know what to do when equipment fails in the middle of rush hour. A variety of talents is therefore desirable, if not essential.
5. *Tie-In with Franchisor's Philosophy and Values.* It is essential that franchisees fully comprehend and agree with the philosophy and values established by the franchisor. The emphasis placed on operational aspects and quality should be fully adhered to. Franchisee-franchisor relationships function smoothly by mutual understanding of these values. Franchising relies heavily on repeat business, which may occur at the franchisee's unit or at others within the franchise system. Franchisors are also interested in the development of image, which will attract investors in addition to consumers. Franchisees have to buy into a certain viewpoint in order to become successful. The likelihood of a franchisee's doing so is judged mainly during personal interviews.
6. *Willingness to Devote Full time to the Operation.* Many franchisors discourage absentee franchisees. To succeed in franchising, full-time devotion of the franchisee is necessary. Not the investment alone, but dedicated day-to-day operation, is desired by franchisors. Full-time involvement also helps the franchisee to become fully aware of operational details as well as to be creative in the development

of the restaurant franchise. With franchisee involvement in mind, many franchisors do not accept absentee or part-time ownership. There are also restrictions on corporations, absentee investors, and partnerships.

7. *Willingness to Relocate.* Restaurant franchises are built in many areas and the willingness of a franchisee to relocate is important. Many franchisees prefer locations close to their place of residence. For new franchisees, it is crucial to be willing to relocate if needed. Willingness to relocate is also an indicator of the interest of the franchisee and his or her belief in the franchise system. Application forms normally include a question about the potential franchisee's willingness to relocate.

8. *Successful Completion of Training.* Training sessions orient franchisees to the concept, operations, and other aspects of the franchise. It is essential for the franchisee to successfully complete the training program. The type of training and other details are provided by the franchisors, and franchise agreements normally state that training constitutes an essential part of the contract. Training can occur pre-opening or continuously. In order to be successful in the training course, the franchisee needs adequate basic knowledge related to business and/or the motivation to learn.

9. *Long-Term Commitment.* Franchising demands a fairly long-term commitment, as franchising agreements normally range from five to twenty years. Franchising calls for dedication to the franchise system in addition to a firm belief in the franchisor's philosophy. People who move from one business to another for short periods of time are usually not suited to franchising.

10. *An Understanding of the Concepts of Franchising.* Prospective franchisees should understand franchising and must be psychologically prepared to accept and agree to the franchisor's role. The pros and cons of franchising should be well understood by the prospective franchisee. The

amount of independence possible in franchising should also be understood clearly by the franchisee. Although franchising is based on selling expertise to others, it is not the place for those who need or want total control.

11. *Willingness to Work with People.* The restaurant business requires working with people at all times. A successful franchisee is willing to work with and for them. Thus, a franchisee has to have good communication skills in addition to leadership skills—the capability to get work done by other people. A franchisee should love working with and for customers, because restaurant operations sell both products and services.

SELECTION OF A FRANCHISOR BY THE FRANCHISEE

Franchisor selection is as important to a franchisee as the selection of a franchisee is to the franchisor. A franchisee is taking a risk in deciding to go with a franchise system. It is not only the financial investment involved, it is the time, effort, and personal commitment that go with the decision. This decision is especially critical for the first-time franchisee. Joining a restaurant franchise system that is well established and has a proven record of success involves less risk but usually is expensive with respect to initial fees and royalties; in addition, such a system usually imposes strict selection criteria. Newer franchises may be less expensive but have greater associated risks. A careful assessment is therefore necessary before a prospective franchisee selects a franchise.

Franchisor Qualifications

Common problems are listed as warning signs of less credible franchisors in Figure 5.1. Described in the following paragraphs are some franchisor aspects that should be considered by a franchisee before making a decision.

- Need cash for ongoing operations
- No company-owned or model units
- Not selective about franchisees
- Fees out of line; either too much or too little
- Pressure to hurry decision
- Too many promises of success
- No history of past successes in business endeavors
- Working out of home
- Listing under (900) numbers (one has to pay certain costs for the telephone call)
- No set training program
- Too high initial fees
- Lack of well-planned operational manual
- Too many unresolved questions
- No list of franchisees available as references
- Too many verbal commitments

FIGURE 5.1   Warning signs of less credible franchisors.

1. *Financial Status of the Franchise.* The franchisor must be reasonably secure financially. Because considerable investment is involved, both franchisees and franchisors should be on sound financial footing to undertake this venture. The financial health of a company can be determined from documents such as the financial statement in the disclosure document, stock market reports, financial analyst's reports, the willingness of franchisor to provide financial data, and the overall investment of the franchisor in the system. If franchisors are working hard on selling franchises alone, just to cover their ongoing operating expenses, then they do not have adequate finances to maintain and develop the franchise system. Start-up costs and projected sales volume should be carefully and critically assessed. A good franchisor is always working on new and better techniques to keep the franchise system profitable.
2. *Selection Procedures for Franchisees.* A franchisor's selection procedure gives a good prognosis of future intentions and

commitment. Because the franchising contract is a long-term commitment, franchisors must be selective when considering franchisees. The selection process should be based on the attributes of franchisees discussed earlier in this chapter rather than on initial investment alone. If the franchisor is desperate for money, chances are that he or she may not be in this franchise business very long. Some of the most successful franchises started with no or very low initial fees but a lot of support and input from the franchisors.

3. *Company-Owned versus Franchised Restaurants.* It is always desirable for the franchisor to have numerous company-owned franchises for the following reasons:
    a. Franchisors should preferably be in the same business as the franchise they are trying to sell. A franchisor who is solely dependent on franchise restaurants selling chicken products is likely to work harder on the development and profitability of restaurants than one whose primary interest is in other types of business. Research and development efforts will be finely tuned toward marketing products and services, thereby increasing sales of restaurants.
    b. The franchisor is aware of the operational aspects of the restaurant and able to solve problems readily.
    c. The franchisor is conscious of the costs involved and works hard to control food and labor costs. These cost-control measures may be passed on to franchisees.
    d. Company-owned restaurants are indicative of a franchisor's considerable investments in land, buildings, leases, and equipment, which make him or her work hard to get optimum return on investment. The franchisor's efforts are focused on marketing and customer satisfaction.
    e. The franchisor has an interest in the growth and success of the restaurants, which may indirectly benefit the franchisees. If franchisors are interested only in selling franchises, operational matters can be neglected.

f. The franchisor has an exact idea of how the franchisees may react to proposed changes. The franchisor has an opportunity to implement those changes in the company-owned units as a pretest or as an example for franchisees. For example, when a new product is introduced it can be test-marketed at one of the company's stores and its profitability demonstrated. If any modifications to the physical attributes of the store are planned, they can be first applied to a company-owned store. This also has a positive impact on the franchisor-franchisee relationship.
g. The franchisors are likely involved in long-term planning for the restaurant franchise rather than dependent solely on selling franchises or on short-term efforts, such as increasing gross sales so that more royalty fees can be collected.
h. A means of comparative assessment can be provided to the franchisees. If something can be done profitably in a company-owned restaurant, the same can be done in a franchised unit.
i. Company-owned stores demonstrate the leadership role of the franchisor and inspire confidence among franchisees, providing incentive for franchisees to equate or excel in their performance.

A debatable question is the ideal ratio of franchised and company-owned restaurants that need to be maintained by a franchisor. A rule of thumb is that 20 to 30 percent of the total restaurants within the system should be company owned. For larger chains with several thousand restaurants, the recommended figure for company-owned restaurants is about four hundred.

4. *Track Record of Franchisor's Success.* The franchisor's success can be assessed from the Federal Trade Commission (FTC) disclosure document or the Uniform Franchise Offering Circular (UFOC). The success of the franchisees within the system also provides a reasonably good indication. One

way to assess success is to look at the unit sales of a franchised restaurant. It is advisable to compare average sales of about five units at different locations. New units should be excluded from these averages, as many of the new franchise restaurants have high gross sales soon after the opening that may level off in time. Inflation can also have an impact on the sales trend, as price increases can misleadingly indicate an increase in sales. An adjustment for impacting external environmental factors should be applied before interpreting the results. It is useful to compare the percentage of consumers within a segment of the population to the total population of that segment. Franchised restaurants cater to the needs of different segments of the population; therefore, comparison of the demographic data is important.

5. *Innovative Attributes of the Franchisor.* Franchising is fundamentally based on innovation and uniqueness in products and services provided. In order to be competitive, a franchisor should always be involved in innovative additions to the products and services offered by the franchise system. These can be in the form of additions to the menu, refinement of equipment, new methods of preparation, or a revised method of service. Successful franchises are continuously upgrading almost every constituent of the franchised business. This need is evidenced from the trends in franchise restaurants. Franchises that started strictly as fast-food operations selling hamburgers are now selling chicken, fish, and steak and have drive-throughs, salad bars, and delivery services. Innovations should be applicable to the franchise system, which demands simplicity and duplication on a large scale.

6. *Staff at the Corporate Headquarters of the Franchisor.* The involvement of the headquarter staff gives a picture of the overall belief of the personnel within the organization in the entire franchising concept. The staff demonstrates the level of cooperation of all departments, including marketing, re-

cruitment, training, and operations, in working closely for the success of the system. Management policies demonstrate the mission and philosophy of the franchisor. The way a franchisee is treated by the staff may indicate how satisfactory future relations will be. It also shows how well the company is organized and run. Because franchisees will be dealing with the staff continuously, all such aspects need consideration when selecting a franchisor.

7. *Support Services Provided by the Franchisor.* The types of support services provided by the franchisor are important to franchisees. These services are outlined in the disclosure documents and are explained to all franchisees. How much these services are needed by and provided to franchisees within the system should be assessed by a prospective franchisee. The proportion of investment in and interest devoted by the franchisor to these services should be an important consideration. From site selection to the opening of the restaurant and the maintenance of quality standards, services provided by the franchisor should be assessed. Emphasis placed by the franchisor on upkeep of the quality of products and services gives an indication of the success of the franchise. It is quality that the consumers want and demand, and so it is important to assess it very carefully. The saying "a chain is as strong as its weakest link" applies to the franchise restaurant. When a franchise restaurant fails to meet standards, the entire system is affected. If this negligence in maintaining quality persists, then the entire franchise system may be in trouble. A franchisor should be firm, fair, and consistent in applying and enforcing standards. Negligence on the part of the franchisor may indicate a lack of long-term interest and incentive for the growth and development of the franchise. In addition to quality control, other basic services provided include training, marketing, and consulting on operational matters.

8. *Franchisor's Responsiveness to the Needs of Franchisees.* A good franchisor realizes the cooperative power of the franchisees in the development of the total franchise system. Franchisees can provide valuable input that helps in the operation and profitability of the restaurants. Franchisee advisor committees (FACs) are formed by some franchisors to obtain input from the franchisees. The franchisor should assume a leadership role in the system and help in making critical decisions, after considering the franchisee's input. One way to assess the franchisor's involvement is to consider the channels of communication established between the franchisor and the franchisee. Communications can be in the form of formal publications, newsletters, meetings, conferences, forums, and electronic bulletin boards. Any new ideas being considered for final adaption should be conveyed first to all franchisees for consideration and discussion. Essentially, there should be ongoing communication between the franchisees and the franchisor. This communication is an indication of the respect shown by the franchisor to the franchisees.

9. *Costs Involved in Franchising.* Payments required by the franchisor should be carefully understood and investigated by a prospective franchisee. Some of these fees include (a) a one-time initial franchise fee, (b) a security deposit, (c) an ongoing monthly royalty fee based on gross sales, and (d) an ongoing advertising fee based on gross sales. Financing required and financing available from private sources and financing institutions should be considered. Financing provided by the franchisor in the form of start-up costs or for future development or restaurant modification should be taken into consideration. Training costs should also be considered. There may be additional costs involved other than those listed by the franchisor, such as leasehold improvements, legal fees, and other start-up costs.

10. *Franchisor's Training Program and Future Assistance.* Training is an important component of franchising. The type and

quality of the training program should be assessed. The duration of the training, type of training, and costs give an indication of the quality of business and the interest of the franchisor. Many restaurant franchisors have well-established facilities where formal training is provided in all operational aspects. McDonald's Hamburger University is an example of such a facility. The operations manual provided by the franchisor provides further insight on future support by the franchisor.

In conclusion, the disclosure documents should be thoroughly reviewed by any prospective franchisee. The location of the market, the type of product and services, and all other aspects should be considered. Legal help should be secured before entering into a franchise agreement because franchising is an important and, in some cases, life-long venture.

Selection of a Franchise by the Franchisee

After consideration of the desirable attributes of the franchisees and the franchisor, the next aspect to be considered is the franchise restaurant itself and if that is the business one should get into. Factors mentioned above should all be considered in the selection of a franchise by a prospective franchisee. Documents should be carefully assessed and the attributes of the franchise evaluated. Some of the questions that should be answered are listed below. Evaluation/checklist forms designed to assess different aspects are shown in Figures 5.2, 5.3, and 5.4, which can be used for objective evaluation.

Self-Evaluation

1. Is franchising the right business for you? Have you considered all the pros and cons of franchising?
2. Do you have the experience required for operating a franchise restaurant?

Check yes (✓); no (x); or (?) on the left-hand side; OR circle the number on the right-hand side. 5 = Excellent; 4 = High; 3 = Average; 2 = Low and 1 = Poor.

___ Do you know the pros and cons of franchising?	1 2 3 4 5
___ Do you have experience in restaurant operations?	1 2 3 4 5
___ Do you have interest and motivation to succeed in the restaurant business?	1 2 3 4 5
___ Are you willing to follow franchisor's procedures in running the restaurant?	1 2 3 4 5
___ Do you have enough time to give to your franchised business?	1 2 3 4 5
___ Do you enjoy working with people?	1 2 3 4 5
___ Do you have good communication skills?	1 2 3 4 5
___ Can you successfully complete franchisor's training program?	1 2 3 4 5
___ Are you well-organized	1 2 3 4 5
___ Can you take financial risk?	1 2 3 4 5
Total "yes" _____	Total Points _____
(A total of nine is excellent)	(90 and above points = excellent)

FIGURE 5.2 Franchisee self-evaluation checklist

3. Do you have an interest in and motivation to succeed in the restaurant business?
4. Are you willing to follow the franchisor's procedures in running all aspects of the restaurant?
5. Will you have time to get involved in a franchise system? Do you have too many other commitments, including those to your family, friends, and relatives?
6. Do you enjoy working for people and with people? Do you have a history of getting along well with your supervisors and subordinates?
7. Do you have good communication skills and can you effectively use feedback?
8. Are you capable of successfully completing the franchisor's training program?

Check yes (✓); no (x) or (?) on the left-hand side; OR circle the number on the right-hand side. 5 = excellent; 4 = High; 3 = Average; 2 = Low and 1 = Poor.

___ Do you like the concept, product, and services of the restaurant?	1 2 3 4 5
___ Do you believe in the franchisor's standards pertaining to Quality, Value, Service, and Cleanliness?	1 2 3 4 5
___ Are you familiar with the menu items and are you confident that those items are popular?	1 2 3 4 5
___ Is the location of the restaurant appropriate for the products and services offered?	1 2 3 4 5
___ Do you feel comfortable working with the processes and procedures outlined by the franchisor?	1 2 3 4 5
___ Do you like the decor, interior design, seating arrangement, drive-thru, and other facilities planned by the franchisor for the restaurant?	1 2 3 4 5
___ Are you satisfied by the average profitability of the restaurant?	1 2 3 4 5
___ Do you feel that all fees payable to the franchisor are reasonable?	1 2 3 4 5
___ Are you confident of having competent personnel in your franchised restaurant?	1 2 3 4 5
___ Does the restaurant have or have a potential for a competitive edge in that location?	1 2 3 4 5
Total "yes" _____	Total Points _____
(A total of nine is excellent)	(90 and above points = excellent)

FIGURE 5.3 Franchise restaurant evaluation checklist

9. Do you plan your activities well in advance and are you well organized in your daily routine activities?
10. Are you self-confident enough to undertake risk in business? Have you successfully overcome failures in the past?

Evaluation of Franchised Restaurant

1. Do you like the restaurant concept, its products and services?
2. Do you believe in the QVSC (Quality, Value, Service, and Cleanliness) standards?

Check yes (✓); no (x); or (?) on the left-hand side; OR circle the number on the right-hand side. 5 = Excellent; 4 = High; 3 = Average; 2 = Low and 1 = Poor.

___ Have you read and understood the disclosure document?	1 2 3 4 5
___ Are you satisfied with the duration of the contract?	1 2 3 4 5
___ Are you comfortable with your responsibilities outlined in the disclosure document?	1 2 3 4 5
___ Are you willing to go along with the franchisor's trademark and proprietary rights?	1 2 3 4 5
___ Do you understand the contents and proper use of the operation's manual?	1 2 3 4 5
___ Do you agree with the advertising policy of the franchisor?	1 2 3 4 5
___ Do you understand the accounting and bookkeeping procedures set by the franchisor?	1 2 3 4 5
___ Do you agree with all conditions set for the termination of the agreement?	1 2 3 4 5
___ Would you understand and agree with the covenant of not competing with the franchisor?	1 2 3 4 5
___ Are you satisfied with the training program and start-up services of the franchisor?	1 2 3 4 5
Total "yes" _____	Total Points _____
(A total of nine is excellent)	(90 or above points = excellent)

FIGURE 5.4 Franchisor evaluation checklist

3. Are you familiar with menu items, and are you confident that they are popular in the area where this restaurant is located?
4. Is the location of the restaurant appropriate for the type of menu being offered? Is there a room for growth and profitability of this restaurant?
5. Do you feel comfortable working in the kitchen and using processes involved in the preparation of all menu items?
6. Do you like the decor, interior design, seating arrangement, drive-in, and other facilities at the restaurant?

7. Are you satisfied with the average profitability of this and/or similar restaurants?
8. Do you feel that all fees charged by the franchisor, such as initial franchise fees, royalty fees, and advertising fees, are reasonable and within your means?
9. Are you sure that you have a competent team of personnel working in the restaurant?
10. Does the restaurant have a competitive edge in that region and will this continue in the foreseeable future?

Evaluation of the Franchisor and Franchising Document

1. Have you carefully read the disclosure document? Have you considered the document with your legal advisor?
2. Are you satisfied with the duration of the contract set in the franchising document?
3. Are you comfortable with your responsibilities as outlined in the disclosure document?
4. Do you have any problems with the proprietary marks and rights of the franchisor?
5. Do you understand the contents and proper use of the operating manual?
6. Do you agree with the advertising policy of the franchisor? Will it satisfy your marketing needs?
7. Do you thoroughly understand the accounting and bookkeeping policies for franchisees set by the franchisor?
8. Do you understand and agree all conditions set by the franchisor related to the termination of the franchising agreement?
9. Do you have any problems with the covenant not to compete without permission from the franchisor?
10. Are you satisfied with the training and start-up assistance provided by the franchisor?

Challenges of Franchising

The International Association of Franchising, in its *Franchise Opportunities Guide,* lists the following challenges of franchising:

Challenge 1: Working Within the System. People who have difficulty following directions or who dislike working within a system find franchising extremely frustrating. Conformity to the franchisor's system is critical if consistency among franchises is to be maintained. However, there are areas, such as marketing, where a franchisee can be creative.

Challenge 2: The Risk. While it is true that purchasing a franchise entails less risk than starting an independent business, risks still exist. Because you own the business, you, to a great extent, determine the success of your venture. The franchisor may have a great program and a respected name, but, in the final analysis, much of the risk is in your hands.

Challenge 3: Working With the Franchisor. Buying a franchise can be closely compared to entering into a marriage. Both are legally binding relationships that can last for a long time. Your relationship with the franchisor and his or her staff is extremely important. Get to know your franchisor through the following methods:
a. Visit the corporate headquarters. Seek to get a feel for the staff and how smoothly the operation runs.
b. Talk to other franchisees. Ask what their relationship with the franchisor is like.
c. Read as much about the franchise as possible.

Challenge 4: False Expectations. Some people enter franchising expecting instant success. Perhaps the reason some expect this is the tremendous success achieved by some franchisees. However, this success did not come without hard work and great effort. Franchising, like any other business, requires tremendous time, initiative, and industry. Obtain from the franchisor as realistic a picture as possible as to what is required in operating that particular franchise.

Challenge 5: Managing the Business. Some individuals are more prepared to manage a business than are others. They have some business experience and have learned to get along well with people. Other individuals may find that managing a franchise is a tremendous burden. You must honestly assess your preparation to run a business. If you find that you have little or no experience, you may want to seek special assistance from the franchisor in business management.

In summary, franchising provides exceptional opportunities for many people. While buying a franchise is not an absolute guarantee that one will be successful, many of the pitfalls of individual ownership can be avoided.

Restaurant Study 7

Churchs Chicken

THE HISTORY OF CHURCHS CHICKEN

In 1997, Churchs Chicken celebrated its 45th year of operation and continues its rich southern history and phenomenal growth.

George W. Church, Sr., of San Antonio, Texas, founded Churchs Chicken in 1952. Church, a retired incubator salesman with more than 20 years in the poultry industry, conceived the idea of offering freshly cooked, quality fried chicken at a time when only hot dogs and ice cream were marketed fast-food style.

Church reasoned that the foodservice industry would have to change its approach in order to capitalize on the opportunities created by population growth and increased mobility. By cutting the frills common to the restaurant industry philosophy of the day, Church felt that he could deliver his product profitably at low cost with a more efficient use of capital and employees.

The first "Churchs Fried Chicken to Go" was located in downtown San Antonio, across the street from the Alamo. The restaurant sold only fried chicken. Church added French fries and jalapeños to the menu in 1955. George Church's idea paid off, and at the time of his death in 1956, four Churchs were open. Other members of the family became active in the business, and by 1962 the chain had grown to eight locations in San Antonio.

George W. "Bill" Church, Jr., assumed chief operating responsibility for the family business in 1962. His father had already proved the economic viability of a low-overhead food outlet serving take-out food at a modest price. Bill Church dreamed of building the business into a nationwide organization. Church and his management team stuck to the basics, and from 1962 to 1965 concentrated on rapid but tightly controlled expansion limited to the San Antonio area. By 1965, Bill Church and his older brother Richard (with whom he worked for about a year) had perfected a marinating formula for Church's Fried Chicken that could be put together almost anywhere in the world. The formula remains a closely guarded secret.

By 1967, the company was set to expand, and less than a year later it established the first Churchs restaurants outside Texas.

The Church family was bought out in October 1968, and in May 1969 Churchs Fried Chicken, Inc., became a publicly held company. At the end of 1969, more than 100 Churchs restaurants were in operation in seven states. Between 1969 and 1974, Churchs grew by an additional 387 restaurants. At year-end 1974, there were 487 Churchs in 22 states, with total revenues of more than $100 million. The highlight of this period was the opening of the national headquarters complex and manufacturing plant on a 6-acre site in northwest San Antonio.

International expansion began in 1979, with the announcement of the first Churchs in Japan. The company subsequently established locations in Puerto Rico, Canada, Malaysia, Mexico, and Taiwan, under the brand name Texas Chicken.

The Churchs concept has always been based on the philosophy of simplicity of operation, starting with a limited menu, equipment designed to do one thing well, people trained to do one thing well, and control of operational costs. From real estate, to restaurant design, to purchasing, Churchs operations are the result of a strong focus on doing the most for the least.

By 1989, Churchs was the second-largest chicken franchise organization in the United States. That was the year it merged with the number-three organization, Popeyes Famous Chicken & Biscuits, headquartered in New Orleans. The Churchs concept remains distinct and separate from Popeyes.

While chicken continues as the mainstay, over the years menu additions have broadened the appeal to Churchs' loyal customers. Churchs also developed a unique honey-butter biscuit after the merger to complement the company's southern-style recipe chicken.

On November 5, 1992, America's Favorite Chicken Company (AFC)—now called AFC Enterprises—officially became the parent company to Churchs Chicken and moved its operations to headquarters in Atlanta. AFC initiated a 100-Day Action Plan and introduced several new programs to revitalize Churchs entire restaurant system in areas of operations, marketing, menus, reimaging, and growth. Churchs continues to improve and expand through aggressive reimaging programs and by seeking alternative development opportunities, including co-branding, convenience stores, grocery stores, and travel centers. Churchs small footprint and simple operating system are ideal for host venues.

As the neighborhood restaurant, Churchs supports the communities in which it does business. In 1992, Churchs committed to build 100 homes around the world with Habitat for Humanity. In addition, Churchs supports the United Negro College Fund (UNCF) and the Hispanic Association of Colleges and Universities (HACU) through annual

140   Restaurant Franchising

(a)

(b)

Churchs Chicken Restaurant (courtesy AFC Enterprises 1998).

scholarships and sponsorship of the UNCF Lou Rawls Parade of Stars Telethon. Churchs also sponsors more than 200 elementary schools around the country through its Adopt-A-School program.

Known for its Southern-style chicken, Churchs also serves southern specialties including okra, coleslaw, mashed potatoes, corn on the cob, and its unique honey-butter biscuits. Churchs Chicken has more than 1,250 locations in 26 states and eight countries worldwide. The 1997 system-wide sales were $720 million.

Restaurant Study 8

Pizzeria Uno

A BRIEF HISTORY OF UNO'S

Uno's began in 1943 when Ike Sewell opened a restaurant at the corner of Ohio and Wabash Streets in Chicago. Ike redefined pizza by creating a deep-dish crust and filling it with huge quantities of the freshest ingredients. Chicago Deep Dish Pizza was born and the rest is legend.

In 1979, after numerous, unsuccessful attempts by others to "take Uno's national," Aaron Spencer, now Chairman of the Board of Uno Restaurant Corporation, succeeded in gaining Ike's agreement to do just that.

Now based in Boston, Uno Restaurant Corporation is a public company whose common stock is traded on the New York Stock Exchange under the Symbol "Uno." Uno Restaurant Corporation, through subsidiaries and affiliates, and Uno's franchisees own and operate more than 151 full-service establishments in the United States, Canada, Puerto Rico, and South Korea.

Over the years, Uno's has evolved from a pizzeria into a full-service casual restaurant specializing in the Original Deep Dish Pizza. The addition of more variety on our menu has been driven by our participation in the extremely competitive Casual Restaurant segment of the industry.

The powerful combination of a unique proprietary product with menu variety, an exciting new dining environment and creative marketing has enabled us to achieve comparable sales figures among the top in the industry.

Pizzeria Uno Corporation has carved out a unique niche in the $20 billion a year casual dining segment of the restaurant industry. They are the only full-service, casual theme restaurant with a brand-name, signature product—Uno's original Chicago Deep Dish Pizza. Their concept is built on a combination of factors:

- The tremendous popularity of their Chicago Deep Dish Pizza.
- A full and varied menu with broad appeal.
- A flair for fun and comfortable decor in a facility that attracts customers of all ages.
- Superior execution achieved through management experience, proven training programs and innovative business systems.

142   Restaurant Franchising

An exterior view of a Pizzeria Uno Restaurant (courtesy Uno Restaurant Corp.).

Franchising is a key element of the strategic vision for the future. They want to work with experienced, multiunit operators who can become part of the Uno team.

They offer an excellent business opportunity based on the equity of their brand name, the quality and popularity of their signature product,

An interior view of a Pizzeria Uno Restaurant (courtesy Uno Restaurant Corp.).

and the proven capabilities of their professional organization. They have the products, the concept, the distribution systems, and the operating procedures to build the Uno concept. By teaming with strong franchisees, they intend to broaden and strengthen their markets to the mutual advantage of Uno, their franchising partners and stockholders.

Uno's key distribution strategy is full-service restaurants. This merchandising approach takes full advantage of Uno's clearly differentiated identity. The freshness of the concept and the frequent introduction of new products, backed by creative promotion and marketing, offer the potential for a satisfying business operation.

We are seeking prospective franchisees with the experience and resources to develop multiple unit operations in new and existing markets.

Uno Restaurant Corporation, the parent and owner of Pizzeria Uno Corporation, is a publicly owned company listed on the New York Stock Exchange.

Uno benefits from clearly defined strategies that have enabled the Company to chart a steady course over the years. These strategies help to:

- Promote brand awareness and acceptance.
- Differentiate their restaurants—from the competition.
- Stake out a niche in the growing casual theme dining segment of the restaurant market.

A key addition to their strategies occurred in 1993/94 with the addition of sauté stations, grills, and fryers in all their kitchens. These new cooking capabilities greatly increased the quality of nonpizza products and have scored with their customers.

Their customer satisfaction since 1994 is largely attributable to their ability to prepare every item on the menu to perfection and with greater efficiency.

Another key to their concept is the popularity of their new prototype restaurant design, which reflects back to their Chicago origins and emphasizes fun, comfort, quality, and high-value dining at very affordable prices.

They use a variety of innovative systems and procedures to help operate efficiently and effectively. Franchisees have access to these systems which Uno has developed and refined over the years.

Some of the tools they provide to franchisees are:

- Recipes
- Prototype design
- Operating manuals and procedures
- Tools for analyzing food and labor costs
- Cash register systems
- Purchasing systems

CHAPTER 6

Franchise Application and Franchise Package

Once a franchise is available for business, potential franchisees start requesting franchise packages and application forms. The initial contact provides the first impression of the franchisor in business dealing. The request is normally made by phone or in a brief letter requesting information. In either case, the response from the franchisor is of vital importance for both parties. Some franchisors have recorded messages where one can leave an address where information may be shipped. In some cases, the inquirer is bluntly told that the franchisor is no longer franchising in the area where he or she is calling from. In spite of the fact that some franchisors get many inquiries, corporate etiquette dictates professional handling of these requests. Most franchisors are well prepared to handle such requests and have carefully planned their franchise packages. A person visiting a franchised restaurant may not be aware of the intricacies of franchising and what the franchisor's goals and objectives are. This chapter describes the application process and the franchise package information, all of which is beneficial to both franchisees and franchisors.

APPLICATION PROCESS

Step 1

The application process starts with an inquiry from a prospective franchisee about a particular franchise restaurant. Franchise availability is normally advertised in newspaper and magazines. Once a request is received, the franchisor mails out a package containing preliminary information about the franchise system. This information widely varies from one franchise to another. Typical information packages contain the following information:

- A letter from the franchise sales administrator, director of franchise development, or other representative of the corporation. This letter introduces the concept and other aspects pertaining to franchising. A letter from a restaurant franchise is shown as an example in Figure 6.1.
- A short history of the restaurant franchise. This may vary from a one-page description to several pages and may include photographs and other pertinent information. The history usually starts with the franchise concept, followed by the introduction of the franchise system and development. This is the initial marketing document and sets the tone for attracting prospective franchisees. Actual opening sentences of such histories are shown below:

Godfather's Pizza
One of the nation's leading pizza restaurant chains . . . Godfather's Pizza was founded in Omaha, Nebraska, in 1973 by Willy Theisen.

Hardee's.

December 18, 1990

Dr. Mahmood Khan
18 Hillcrest Hall
Blacksburg, VA 24061

Dear Dr. Khan:

Thank you for your interest in Hardee's franchise program. We are enclosing our standard information that hopefully will answer questions we are asked most often.

At the present time, we are accepting franchise applications in a limited number of geographical areas. Our present base of existing franchisees is sufficient to meet our current development plans in all other areas.

We encourage you to compare the benefits of a Hardee's franchise against our competitors and consult your banker or investment counselor. After such an investigation, we feel that an investment with us could be extremely rewarding. We do encourage you to submit a completed application. Please see our enclosed instructions for doing so. Upon receipt of a completed application, we will review it and, should you meet our minimum financial qualifications, arrange a discussion of opportunities suited to your personal goals and qualifications.

Our minimum financial requirements are a net worth of Five Hundred Thousand Dollars ($500,000), exclusive of your personal residence, with One Hundred and Fifty Thousand Dollars ($150,000) in liquid assets.

We look forward to hearing from you again and should you have any questions, please do not hesitate to call.

Sincerely,

HARDEE'S FOOD SYSTEMS, INC.

Franchise Sales Administrator

Enclosure

HARDEE'S FOOD SYSTEMS, INC. • 1233 HARDEE'S BOULEVARD • P. O. BOX 1619 • ROCKY MOUNT, NC 27802-1619
919/977-2000

FIGURE 6.1 An example of a letter from a franchisor in response to a request from a prospective franchisee (Courtesy Hardee's Food Systems, Inc.)

Specializing in high-quality, thick-crust pizza, Theisen recognized the potential of his product and the importance of expansion, and the first Godfather's Pizza franchise was opened in 1974.

Hardee's
The Hardee's name is well known throughout the franchise and restaurant industry as an established quality-oriented company. It has grown from a single unit in Greenville, North Carolina, in 1960 . . .

Our solid management team is committed to establishing Hardee's throughout the country by means of a well-planned marketing philosophy that includes individuals, or groups, who have the desire to be their own boss and who wish to further the free enterprise system.

Hardee's is a proven method of operation and a value to its present and future franchisees.

- A description of the type of menu offered by the franchisor for use in restaurant. This may vary from a listing to a detailed description. An example of such a menu listing from Hardee's is shown in Figure 6.2.
- Availability of areas for franchisees are also described. Well-established franchises normally do not have many available territorial choices.
- A brief description of the training offered by the franchisor to the franchisees.
- A brief description of other services offered by the franchisors, such as:

site selection
feasibility study
operations and field services
purchasing
marketing
research and product development

HARDEE'S STANDARD MENU

Biscuit Breakfast
Pancakes
Hash Rounds™ Potatoes
Hamburger
Cheeseburger
Big Twin® Double Burger
¼ Pound* Cheeseburger
Bacon Cheeseburger
Big Deluxe™ Burger
Fisherman's Fillet™ Sandwich
Roast Beef Sandwich
Big Roast Beef™ Sandwich
Grilled Chicken Breast Sandwich
Chicken Fillet Sandwich
All Beef Hot Dog
Garden Fresh Salads
French Fries
Crispy Curls™ Potatoes
Apple Turnover
Big Cookie™ Dessert
Cool Twist™ Cones & Sundaes
Soft Drinks & Iced Tea
Milk Shakes
Coffee
Orange Juice
Milk

*pre-cooked weight

NOTE: The above menu items are standard for company owned and operated restaurants as of May, 1990. Some menu items are optional for franchisees due to geographical consumer preferences. This list is subject to change from time to time.

FIGURE 6.2 A page describing a menu included in a franchise package (Courtesy Hardee's Food Systems, Inc.).

building and equipment
opening assistance
accounting
controls
operational support

- A summary of costs required at the initial stages, which normally include:

license fee
approximate total cost of equipment and signs
construction costs
real estate costs
working capital
site improvement costs
training costs
leases
leasehold improvement costs

- The code of ethics for franchisors published by the International Franchise Association is also included in some franchise packages, particularly by those who are members of the association. The code of ethics focuses on the franchisor's obligation to franchisees in all dealings. The code of ethics is shown in Figure 6.3.
- An application form. Normally this is a preliminary application form used for screening by the franchisor. Confidential information that is pertinent to franchising is solicited through this application form. Two examples of application forms are presented in Figures 6.4 and 6.5. In addition to demographic information, this application form requires a complete personal financial statement. A summary of assets and liabilities is required from the prospective franchisee. A signed statement by the applicant is required, and rights to check credit history and verification of the financial history are secured. The major questions in a typical application form are those shown in Figures 6.4 and 6.5.

INTERNATIONAL FRANCHISE ASSOCIATION
CODE OF ETHICS

1. No member shall offer, sell or promote the sale of any franchise, product or service by means of any explicit or implied representation which is likely to have a tendency to deceive or mislead prospective purchasers of such franchise, product or service.

2. No member shall imitate the trademark, trade name, corporate name, slogan, or other mark of identification of another business in any manner or form that would have the tendency or capacity to mislead or deceive.

3. The pyramid or chain distribution system is inimical to prospective investors and to the franchise system of distribution and no member shall engage in any form of pyramid or chain distribution.

4. An advertisement, considered in its totality, shall be free from ambiguity and, in whatever form presented, must be considered in its entirety and as it would be read and understood by those to whom directed.

5. All advertisements shall comply, in letter and spirit, with all applicable rules, regulations, directives, guides and laws promulgated by any governmental body or agency having jurisdiction.

6. An advertisement containing or making reference, directly or indirectly, to performance records, figures or data respecting income or earnings of franchisees shall be factual, and, if necessary to avoid deception, accurately qualified as to geographical area and time periods covered.

7. An advertisement containing information or making reference to the investment requirements of a franchise shall be as detailed as necessary to avoid being misleading in any way and shall be specific with respect to whether the stated amount(s) is a partial or the full cost of the franchise, the items paid for by the stated amount(s), financing requirements and other related costs.

8. Full and accurate written disclosure of all information considered material to the franchise relationship shall be given to prospective franchisees a reasonable time prior to the execution of any binding document and members shall otherwise fully comply with Federal and state laws requiring advance disclosure of information to prospective franchisees.

9. All matters material to the franchise relationship shall be contained in one or more written agreements, which shall clearly set forth the terms of the relationship and the respective rights and obligations of the parties.

10. A franchisor shall select and accept only those franchisees who, upon reasonable investigation, appear to possess the basic skills, education, personal qualities, and financial resources adequate to perform and fulfill the needs and requirements of the franchise. There shall be no discrimination based on race, color, religion, national origin or sex.

11. The franchisor shall encourage and/or provide training designed to help franchisees improve their abilities to conduct their franchises.

12. A franchisor shall provide reasonable guidance and supervision over the business activities of franchisees for the purpose of safeguarding the public interest and of maintaining the integrity of the entire franchise system for the benefit of all parties having an interest in it.

13. Fairness shall characterize all dealings between a franchisor and its franchisees. To the extent reasonably appropriate under the circumstances, a franchisor shall give notice to its franchisee of any contractual breach and grant reasonable time to remedy default.

14. A franchisor should be conveniently accessible and responsive to communications from franchisees, and provide a mechanism by which ideas may be exchanged and areas of concern discussed for the purpose of improving mutual understanding and reaffirming mutuality of interest.

15. A franchisor shall make every effort to resolve complaints, grievances and disputes with its franchisees with good faith and good will through fair and reasonable direct communication and negotiation. Failing this, consideration should be given to mediation or arbitration.

© 1978 International Franchise Association

FIGURE 6.3 Code of Ethics (Courtesy International Franchise Association)

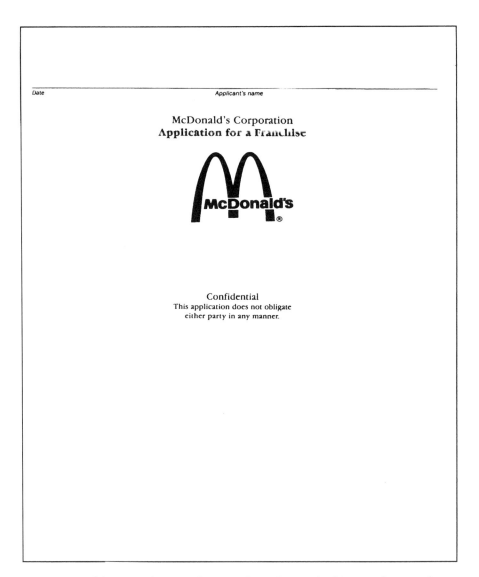

FIGURE 6.4 Franchise application form (Printed with permission of McDonald's Corporation, 1991)

Name	Home phone
Residence	Business phone
City	May we contact you at your business phone?
State/Zip code	Social security number

Previous addresses for past ten years

PERSONAL INFORMATION

Date of birth	Height	Weight	Marital Status
Spouse's name	Names and ages of children		Total dependents

Date and purpose of last physical exam

Describe any physical disabilities or limitations

Have you ever been convicted of anything other than minor traffic violations? Has any judgment ever been entered against you or your company or your employer where you were one of the litigants? Are you involved in pending litigation? If yes, explain.

Of which country are you a citizen?

EDUCATION

Last year of school completed	Name of college and/or postgraduate school	Degree

Describe any training in sales, management or retailing

BUSINESS EXPERIENCE

Present occupation	Position	Dates employed
Company	Address	

Describe duties, number of employees supervised and responsibilities

Previous business experience (Give exact names, address and dates. List most recent first.)

Dates employed	Position	Company	Type of business
Address		Name of supervisor	Reason left
Responsibilities			
Dates employed	Position	Company	Type of business
Address		Name of supervisor	Reason left
Responsibilities			

FIGURE 6.4 *(Continued)*

Page 2 of 5

Dates employed	Position	Company	Type of business
Address		Name of supervisor	Reason left
Responsibilities			
Dates employed	Position	Company	Type of business
Address		Name of supervisor	Reason left
Responsibilities			

Have you ever been self-employed? If so, explain.

Have you ever had a business failure? If so, explain.

PERSONAL FINANCIAL STATEMENT

	$
Salary, wages	
Bonus, commissions	
Dividends, interest	
Real estate income	
Business profits	
Notes/accounts receivable	
Other income—specify source, e.g. trust, spouse, etc.	
TOTAL	$

Please provide details on the following asset verification schedules *(schedule numbers in parentheses)*.

Assets		Liabilities	
Cash on hand and in banks	$	Notes/loans payable to banks *(4)*	$
Vested profit sharing		Notes/loans payable to friends, relatives *(4)*	
Securities *(1)*		Accounts and bills due *(4)*	
Bonds/debentures *(2)*		Real estate mortgages *(7)*	
Notes, accounts and mortgages receivable *(3)*		Other debts or obligations *(6)*	
Real estate—current market value *(7)*			
Net value of business interests *(8)*		**Total liabilities**	$
Other—automobiles and other personal property, etc. *(5)*		**Net worth**	$
Total assets	$	**Total liabilities and net worth**	$

Give names of banks or finance companies where accounts are carried or where credit can be obtained or verified.

Name	Address	Highest extended credit	Purpose

Can you personally meet the McDonald's financial requirements? From what sources?

FIGURE 6.4 *(Continued)*

ASSET VERIFICATION SCHEDULES

(1) Listed securities

No. shares	Description	Pledged yes/no	Current mkt. value
Total			$

(2) Bonds/debentures

No.	Description	Pledged yes/no	Face value	Current mkt. value
Total				$

(3) Notes/accounts/mortgages receivable

Debtor	Relation to applicant	Nature of debt	Maturity date	Original face value	Monthly payment	Present balance
Total						$

(4) Loans/notes/accounts payable (excluding mortgages)

Lender	Relation to applicant	Nature of debt	Secured yes/no	Maturity date	Original face value	Monthly payments	Interest rate	Present balance
Total								$

(5) Other Assets
(e.g.: Stock options, cash value of life insurance, automobiles and other personal property, etc.)

Description	Current fair mkt. value
Total	$

(6) Other debts and liabilities
(e.g.: Insurance loans, alimony, child support, leases, contracts, legal claims, judgments, chattel mortgages, taxes, comaker or guarantor, etc.)

Obligee	Description	Amount
Total		$

FIGURE 6.4 *(Continued)*

ASSET VERIFICATION SCHEDULES—Continued

(7) Real estate

Address and description of property (residential, rental, vacant)	Date acquired	Title in name(s) of	Original cost	Original mortgage amount	Mo. payments (incl. taxes, assessments)	Current market value	Current mortgage balance	Net value
Total						$	$	$

(8) Business interests

Business Address

Name of business	Description	Type (partner, corp., sole)	Names of all owners	Relation to applicant	Percent equity	Buy/sell agreement yes/no	Valuation method (book, earnings multiple, appraisal, agreed value)	Net value your interest
Total								$

Does your spouse or another person have any interest in any of the above assets? If yes, please explain and list assets.

Have any of the above assets been acquired by you as a gift? If yes, specify assets.

From whom?   When?

Personal references (other than employers or relatives)

Name in full	Address	Occupation	Years known
Name in full	Address	Occupation	Years known
Name in full	Address	Occupation	Years known

FIGURE 6.4   *(Continued)*

List any hobbies, community activities, special interests or other pertinent information

_____
_____
_____
_____
_____
_____

Are you related by blood or marriage to any officer of McDonald's Corporation?   Name   Relationship

Are you or your employer providing products, goods or services to McDonald's or franchisees of McDonald's? If yes, please attach detailed information.

Will you devote your full time to this business?

Willingness to relocate is required. In which general geographic area(s) are you most interested (i.e. Northeast, Southeast, Midwest, West)?

Have you ever worked in a McDonald's restaurant? If so, where and when?

Have you ever applied for a McDonald's franchise? If so, where and when?

"I submit the foregoing information as my complete and true personal and financial condition as of the date shown below. In accordance with the Privacy Act (5 U.S. C552a), Freedom of Information Act and The Fair Credit Reporting Act, I expressly authorize any past or present employer, any law enforcement agency, federal, state or local, or any person who has personal knowledge of my character, work experience or criminal records to release this information to McDonald's Corporation. If requested by McDonald's Corporation, I agree to supply statements from my professional advisors (i.e. banker, broker, accountant or attorney) verifying the above assets, and I also agree to furnish copies of Federal Income Tax Returns as filed for the last five years. I understand that McDonald's Corporation is relying upon all the above information as a material factor in considering my application to become a McDonald's franchisee, and I release all persons from liability as a result of true, accurate information.

Signature (Applicant)   Date

Signature (Spouse)   Date

FIGURE 6.4 *(Continued)*

158   *Restaurant Franchising*

CONFIDENTIAL
FRANCHISE APPLICATION
FOR

Applicant's Last Name	First	Middle	Date of Birth	Home Telephone Number ( )	
Residence Address		City & State			Zip Code
Employed By		Position		Business Telephone Number ( )	
Business Address		City & State			Zip Code

APPLICATION MUST BE COMPLETED **IN FULL** BEFORE SUBMISSION TO:
HARDEE'S FOOD SYSTEMS, INC.
ATTN: Franchise Development Department
1233 Hardee's BouleVard
Post Office Box 1619
Rocky Mount, North Carolina 27802-1619
(919) 977-8901 or (919) 977-8821

FIGURE 6.5   Franchise application form (Courtesy Hardee's Food Systems, Inc.)

Franchise Application and Franchise Package

Marital Status _____ No. Dependents _____ Ages _____

Condition of Health _____

Education Completed: High School   1   2   3   4   College   1   2   3   4   5   6   Degree _____

Have you ever failed in a business venture? _____

If yes—When, where and circumstances _____

_____

_____

_____

Have you any outstanding liabilities as a result of such previous failures? _____

If yes, give particulars _____

Will the HARDEE'S RESTAURANT be owned and operated by yourself or a group? _____

Please explain fully: _____

_____

_____

Operating Partner _____ % Ownership _____

Amount of capital available for this business _____

Describe fully _____

_____

_____

Are you presently connected in any way with the Fast Food Restaurant industry? _____

If so, in what capacity? _____

_____

_____

Territory in which you are applying for a franchise: State _____ County _____

Would you consider any other area? _____ What area? _____

How did you hear about HARDEE'S FRANCHISE? _____

         ☐ Paper _____

         ☐ Magazine _____

         ☐ Other _____

NOTE: This is not a contract and this form incurs no obligation on either party.

Page Two

FIGURE 6.5   *(Continued)*

Character References:	Address	City & State	Zip Code
1.			
2.			
3.			

Employment Experience:

From	To	Firm Name and Address	Position	Annual Income

Business Investments (List all Businesses in which you have Financial Interest)

Name of Business & Address	Position	Yr. Started	Income

Credit References (At Least 5)

Name of Company	Address	City & State	Zip Code
(1)			
(2)			
(3)			
(4)			
(5)			

Business References (At Least 5)

Name of Individual	Address	City & State	Zip Code
(1)			
(2)			
(3)			
(4)			
(5)			

Attach Additional Sheets If More Space Is Needed For The Above

FIGURE 6.5 *(Continued)*

Personal Banking References (Also List Savings & Loan, Money Market Funds & Other Financial Institutions)

Name of Bank or Institution	Address	City & State	Zip Code

PERSONAL FINANCIAL STATEMENT AS OF   /   /

Assets	Liabilities
Cash On Hand and In Bank (Itemize in Schedule A)	Notes Payable to Bank(s) (Itemize in Schedule H)
Mortgages or Notes Due To Me (Itemize in Schedule B)	Loans Against Cash Value of Life Insurance
Other Notes/Accounts Due To Me (Itemize in Schedule C)	Notes Payable To Others (Itemize in Schedule H)
Stocks and Bonds (Itemize in Schedule D)	Charge Acct./Bill Payable (Itemize if More Than $5,000)
Cash Value of Life Insurance (Itemize in Schedule E)	Real Estate Mortgage Payable (Itemize in Schedule F)
Real Estate Owned (Itemize in Schedule F)	Taxes Due (Itemize in Schedule I)
Equipment/Office Furniture (Itemize in Schedule G)	Other Liabilities (Itemize Below)
Household Furnishings and Other Personal Property	
Other Assets (Automobile(s) and Business Venture(s))	
	Total Liabilities (Add all of the above)
	Net Worth (Assets Minus Liabilities)
Total Assets (Add all of the above)	Total Liabilities and Net Worth

---

Income Statement For The Year Ending _____ , 19 _____

- Earned Income (Salary, Fees, Etc.) _____
- Bonus (Commissions) _____
- Interest (Dividends) _____
- Rental Income _____
- Other Income (List) _____
- Total Income (Add all of the above) _____

Page Four

FIGURE 6.5 *(Continued)*

Schedule A	Cash On Hand and In Bank		
Name of Bank	City & State	Type of Account	Balance

Schedule B	Mortgages or Notes Due To Me		
Address of Property	Maker of Mortgage or Note	City & State	Balance

Schedule C	Other Notes or Accounts Due To Me		
Maker of Note	Description of Note	City & State	Balance

Schedule D	Stocks and Bonds (Attach Copies or Statements)		
Name of Company	Number of Shares	Market Value	Amount Pledged and To Whom

Schedule E	Cash Value of Life Insurance		
Name of Insurance Company	City & State	Face Amount	Cash Value

FIGURE 6.5 (Continued)

Schedule F Real Estate Owned

Description of Property	Name on Title	Cost	Market Val.	Bal. Owed	Mortgage Holder

Schedule G Equipment or Office Furniture

Description of Property	Name of Owner	Cost	Market Val.	Bal. Owed	Mortgage Holder

Schedule H Notes Payable to Banks and Others

Name of Note Holder	City and State	Orig. Bal.	Bal. Owed	Terms

Schedule I Taxes Due

Type of Tax	Amount Owed	Date Due	Unpaid from Prior Year

The information contained in this statement is provided for the purpose of obtaining, or maintaining credit with you on behalf of the undersigned, or persons, firms or corporations in whose behalf the undersigned may whether severally or jointly with others, execute a guaranty in your favor. The undersigned understands that you are relying on the information provided herein (including the designation made as to ownership of property) in deciding to grant or continue credit. The undersigned represents and warrants that the information provided is true and complete and that you may consider this statement as continuing to be true and correct until a written notice of a change is given to you by the undersigned. You are authorized to make all inquiries you deem necessary to verify the accuracy of the statements made herein, and to determine my creditworthiness. You are authorized to answer questions about your credit experience with me.

Signature (Ink) _____ Date _____

FIGURE 6.5 *(Continued)*

A prospective franchisee should carefully review all the information included in the package. Special attention should be paid to franchise availability and to costs involved in obtaining a franchise.

Step 2

If further interested, the prospective franchisee, after reviewing the entire application package, submits a basic franchisee application form. Submitting this application does not obligate either the applicant to accept nor the franchisor to offer a franchise. This application is reviewed either by the regional licensing manager or an assigned officer in the franchising department of that particular franchise system. The application form is used for preliminary screening and provides for a review of the potential of a franchisee to meet all the requirements of the franchise operation. All qualifications desired in a franchisee, particularly financial net worth, are assessed. Factors such as creditworthiness and willingness to relocate are carefully reviewed in addition to the experience of the franchisee. If the application survives the preliminary screening, a company sales representative may contact the prospective franchisee. This contact may be for obtaining more information or clarification of the information provided in the application form. This communication provides an opportunity for the prospective franchisee to ask questions related to the franchise. The franchisor also assesses the interest of the franchisee in pursuing the application.

Step 3

In the presence of continuing mutual interest, a representative calls all promising candidates to make arrangements to furnish legally required information and, possibly, arrange for an interview. This interview is designed to include a discussion of the philosophy and goals of the franchisor in addition to assessment of the franchisee's background information, financial status, and genuine interest in the franchise. Often this is the first face-to-face meeting between franchisor and franchisee. It provides for a frank and open exchange of information and assessment of the potential of a long-term relationship between the two parties. The interview may be conducted over the phone or may take place at one of the regional offices of the franchising corporation.

Step 4

If there is mutual approval and the franchisee meets the minimum financial criteria for net worth and liquidity, he or she may be invited to the corporate headquarters or area offices to continue discussion with the staff. At this point a complete review of the franchise program is conducted. Several personal interviews with individuals in different areas may be arranged. At the end of these meetings, normally the decision of approval or disapproval is conveyed to the prospective franchisees.

Step 5

If approved, further discussion concerning the geographic location of the restaurant, site development, lease requirements, leasehold development, and renting or buying options takes place. The candidate may be assessed on practical capability, which is normally done by on-the-job training. Usually this is provided at a local restaurant of the franchise and is brief. If the prospective franchisee already has experience in that particular franchise, the training requirement may be waived or reduced in duration. This on-the-job training allows both the franchisor and the franchisee to evaluate operations in the restaurant environment and provides the prospective franchisee with a closer look at the concept and operation of the franchise.

Step 6

After a location is secured by the franchisee by lease, rent, or purchase, and approved by the franchisor, the franchisee and/or representative(s) undergoes an intensive training program in operations methods. This formal training provides the complete knowledge and hands-on experience related to operations and management of the franchise. The type and duration of this training varies with the franchise.

After training, the franchisee is ready to sign formal documents and start construction or operation of the franchise restaurant. If a franchise is not available at the time or at the location desired, an approved franchisee may have to undergo a waiting period, which may be lengthy for popular franchises.

OPERATIONS PACKAGES AND MANUALS

Operations packages sent to prospective franchisees before they sign a franchise agreement are normally recruiting tools, providing general information necessary for the franchisee. After completion of the franchise agreement, a set of confidential franchise packages, consisting of several manuals, are provided by the franchisor. These manuals are basic tools focusing on all aspects of the franchise. The success of the entire franchise operation depends on understanding and utilizing these manuals. These must therefore be carefully prepared by the franchisors and written at a level that can be easily comprehended. Major manuals and information packages are described below.

Operating Manual

The operating manual is the most referred to, key manual of a franchise restaurant. Because it outlines all operational aspect of the franchise, the material is extremely confidential. All major functions within the restaurant are addressed in this manual, which must be clearly written and convey all information in a sequential and systematic manner. It should also be updated and revised periodically.

Its applicability should be tested continually and areas that are not clear should be rewritten. Often franchisees and questions posed by them provide good feedback concerning the usefulness of the operating manual. Illustrations, graphs, figures, and tables should be provided wherever necessary. Sections outlining administrative, functional, and legal aspects should be clearly identified. Many franchises use three-ring loose-leaf binders for the operations manual, thereby facilitating easy addition and removal of sections as and when necessary. Operations manuals should be comprehensive enough to answer all questions related to the operations of a franchise. Operation manuals are the sole property of the franchisors and must be treated confidentially because they outline in detail the basic operations of the franchise system. A sample outline of an operations manual of a franchised restaurant is shown below:

Section 1: Introduction
- brief description of the concept
- history of the franchise
- landmarks and significant changes with time
- franchisor's philosophy
- organizational chart and administrative staff
- franchisee/franchisor responsibilities

Section 2: General Rules
- critical points
- quality expectations
- customer relations
- procedures and policies
- controls and inspections
- warranties
- maintenance requirements

Section 3: Menus and Menu Plans
- menu details
- menu changes

- computerized menu controls
- menu design and display
- mechanics of menu planning

Section 4: Equipment Use and Care
- types of equipment
- equipment use and precautions
- equipment cleaning
- equipment maintenance

Section 5: Food Purchasing
- supplier list
- commodities required
- specifications
- planning purchasing
- buying methods
- regulatory agencies and control
- ethics in buying
- purchase of specific commodities

Section 6: Receiving and Storage
- mechanism of receiving
- storage requirements
- issuing supplies
- inventory procedures
- inventory controls
- inventory levels

Section 7: Sanitation Requirements
- foodborne illnesses and precautions
- safe handling of foods
- safe temperatures and critical points
- personal hygiene and procedures

Section 8: Food Preparation
- procedures for preparing different items
- critical control points
- time and temperature controls

- precautions in preparation
- quality of the products
- production forecasting and planning
- Production sheets
- leftover control

Section 9: Delivery and Service of Foods
- portion control methods
- consumer aspects
- dress code and personal appearance
- service procedures
- delivery procedures

Section 10: Financial Aspects
- cost sheets
- financial statements
- food costing
- labor costing
- profit and loss statement
- balance sheets
- other financial controls

Section 11: Management
- employee and supervisor relations
- recruitment and selection procedures
- performance appraisal
- employee rules, grievance procedures, and turnover
- motivation and job enrichment
- control of human resources
- stress management

Section 12: Restaurant Operation
- restaurant opening and closing procedures
- opening and closing registers
- maintenance and housekeeping
- handling consumer complaints and special requests
- lighting, signage, and atmosphere control

Section 13: Marketing
- advertising policies and procedures
- promotional activities
- community programs
- long- and short-term marketing plans

Section 14: Maintenance
- utilities control
- fire protection
- pest control
- garbage disposal
- parking and drive-in maintenance
- electrical, mechanical, construction, and plumbing
- heating, ventilation, and air-conditioning
- alarms, locks, and security
- music control
- repairs and maintenance
- computer problems and control

The above-mentioned topics can be included in a typical operations manual for a restaurant and modified based on the products and services provided. The manual is an essential component of the services provided by the franchisor. Franchisees consider the manual a reflection of the services provided by the franchisor. It should be used daily by the franchisees while operating a restaurant.

Training Manual

Training is one of the major services provided by a franchisor. The training manual should be prepared carefully and tailored to the educational level of the franchisees. Franchise training programs normally have three components, for which the contents of the training manual are described as follows:

- *Hands-on training programs* are designed to provide practical experience in the day-to-day operation of the franchise. The trainees go through the operational manual and follow its sequence. Usually this training can best be provided in a restaurant in operation. The primary objective of hands-on training is to provide a rotation in each operational aspect of the restaurant. This training is usually supervised by a senior employee within the franchising operation.
- *Formal training programs* are more comprehensive and involve intensive training of the franchisees. All functions of the restaurant's operation and management are included. Formal training sites and training methods are used. An example of a venue for such training is McDonald's Hamburger University. Located in suburban Chicago, the university has well-equipped training facilities and is ranked as one of the only facilities of its kind in the world. Courses offered are accredited by the American Council of Education and earn college credits. Although many restaurant franchisors have their own well-planned facilities, some small franchisors utilize regional offices or local colleges to provide training and course work. Frequent evaluations of the learning resources and course content should be conducted in order to assess the appropriateness of the training program. Successful completion of the training program is one of the prerequisites imposed by franchisors, which requires that there be effective means for evaluating the success of the franchisees. Evaluation can be done by a well-selected outcome assessment, such as the performance of the franchisee when compared to other franchisees within the system.
- *Ongoing training programs* are provided on a continual basis to the staff of the franchisee, often on site or at corporate headquarters. Training manuals for this purpose primarily include the operational aspects of the restaurant. This training can be provided by the franchisee or his or her representative.

Marketing Manual

The marketing manual describes the marketing philosophy of the franchisor and outlines the procedures for marketing products and services. Consumer information, nutritional labeling information, essential ingredient information, and image-building of the franchise are discussed. Ways of securing a market niche for the products and services offered are outlined for the benefit of the franchisees. This manual should also discuss the latest marketing techniques and innovative ways to maintain a market niche in a competitive environment. New product development, marketing strategies, price relationship, and pricing policies are included.

Advertising Manual

Advertising, promotion, signage, and public relations are all outlined in the advertising manual. Promotional materials and policies of the franchisors are described. Plans to advertise using different media are explained. The manual also contains illustrations of materials and examples from past promotions. This manual is updated continually to provide current advertising and promotional information. Included in the manual are promotional activities that can be conducted for the local community and organizations such as scout troops and sports teams.

Field Support Manual

The field support manual is a listing of field support services provided by the franchisor. Support services help to maintain healthy franchisor-franchisee relationships. The manual should identify and outline all services, with names, addresses, and phone numbers of those who should be contacted for help. Services include inspection, recordkeeping, quality control standards, and other controls. Possible approaches to major problems and ways to get feedback from the franchisor are listed in the manual. Field support staff and representatives also maintain contact with the franchisees.

Quality Control Manual

Controlling quality is a major emphasis of any franchisor. Quality pertains both to the food and services offered. The quality control manual contains a description of the quality control measures a franchisor requires to maintain standards, technical controls, maintenance services, leftover control, and handling of consumer complaints. Evaluation procedures and forms used by the inspection staff to assess quality control are also included in this manual. Sometimes these forms are included in another manual referred to as the *site inspection manual.*

There may be other manuals based on the type of franchise. Reports and reporting procedures may be included in a reporting manual. Some franchisors condense some or all of these manuals into one large or two or three smaller manuals. Well-organized and well-written manuals lead to smooth operations and good franchisee-franchisor relationships.

Restaurant Study 9

Captain D's

History

Captain D's was founded in Nashville, Tennessee, in 1969 as Mr. D's and has since experienced rapid growth in 22 markets across the United States.

In 1974, the name was officially changed to Captain D's.

Captain D's operates 365 company-owned and 211 franchise restaurants.

Captain D's is committed to continuing the growth of its franchise system by granting franchise opportunities and providing support for restaurant development to those who are qualified and dedicated to establishing true neighborhood seafood restaurants.

Captain D's is part of the Shoney's, Inc., family. Their history demonstrates the commitment and contributions of all of their franchisees and they look forward to nurturing these relationships for a successful future.

Critical Success Factors

- Captain D's provides quality seafood at competitive prices in an attractive but quick-service atmosphere.
- Their customers demand a variety of seafood, prepared a number of ways, in a short amount of time.
- They have expanded to include baked and broiled items, various types of seafood and side items as well as their staple fried items to continue meeting their customers' wants and needs.
- Captain D's combines the best in purchasing, food manufacturing, and supply and distribution services to provide their restaurants with the finest food available every day.
- Captain D's has a Franchise Advisory Council that provides input in the areas of marketing, food, equipment, and building remodeling. Both franchisees and company employees are involved in the decision making for future D's programs, promotions, and projects.
- Service, quality, value, variety, and atmosphere all contribute to keeping Captain D's a leader in the quick-service seafood market.

Captain D's plans to retain the typical D's customer who enjoys a fried seafood dinner, but also add to the mix another type of customer who wants a more upscale menu variety. The "Coastal Classics" menu ranges from $4.99 to $6.99 on items such as broiled orange roughy, Pacific Northwest salmon, Southern-style catfish, shrimp scampi, and oysters.

(a)

(b)

An exterior view of a Captains D's Restaurant (photo by Vando Rogers)

The year 1997 showed a decline in comparable restaurant sales from previous years, because of increased competition and less effective advertising. This year Captain D's has introduced a new plan. In conjunction with the new ad campaign ("Captain D's Please!"), the Coastal Classics menu and a shift from the traditional sea blue to a cranberry-type seaberry exterior paint color—Captain D's is making 1998 a strong year for expanding our own niche.

Franchising is a positive expansion move for companies that are dedicated to offering their products, services, and commitment to quality to a larger portion of the population in a number of regions/markets. When company concepts and philosophies are introduced to franchisees, these entrepreneurs take the company ideals, along with their individual ideas, and produce amazing accomplishments.

Their franchisees have proven to be excellent members of the Captain D's family. As they face the future, they hope to incorporate more franchisees in more areas of the country, increasing their knowledge and innovation, as well as providing more opportunities to serve their customers.

Restaurant Study 10

Subway

Did you ever wonder how the Subway restaurant chain became the world's largest submarine sandwich franchise, with thousands of independently owned and operated restaurants spanning the globe?

The story begins in Bridgeport, Connecticut, during the summer of 1965. Ambitious 17-year-old high school graduate, Fred DeLuca, was working at a local hardware store trying to earn enough money to pay his college tuition. He was determined to find a way to supplement the minimum hourly wage he was paid. The solution came at a backyard barbecue, during a conversation with his family friend, Dr. Peter Buck.

Dr. Buck, a nuclear physicist, suggested to young DeLuca that he open a submarine sandwich shop. Dr. Buck pointed out that there was a successful sandwich shop in his hometown where everyone, including himself, enjoyed the sandwiches.

With a $1,000 loan from Dr. Buck, a partnership was formed. Pete's Super Submarine opened on August 28, 1965, in a remote location in Bridgeport, Connecticut, not far from the hardware store.

In the early days, the company's weekly meetings took place in the DeLuca family kitchen. It was there, over bowls of homemade pasta, that the partners focused on ways to increase sales and meet the challenges that they faced.

The first year was not an easy one for the young entrepreneurs. At the end of that year, they were not sure if they should proceed with the business. But they endured. Upon opening their second location a year later, the two men realized that marketing and visibility were going to be key factors to the success of their restaurants.

Although they had two sub shops open, Fred and Dr. Buck did not consider themselves successful small business owners. At this point, most of us would have thrown in the towel, but Fred DeLuca and Dr. Peter Buck decided to open a third shop. This time they opened a highly visible location, and it appeared that three was their lucky number. The third store started making money and is still serving sandwiches today.

Additional changes to help increase visibility included shortening the name from Pete's Subs to Subway and introducing the now familiar bright yellow logo.

Their next step was to formulate a business plan that outlined the company goals. In an effort to reach these goals, Subway restaurants began franchising, giving others the opportunity to succeed in their own business venture. The first Subway franchise opened in Wallingford, Connecticut, in 1974.

A decade later, the first international Subway restaurant opened its doors on the island country of Bahrain, off the coast of Saudi Arabia.

In August 1995, the Subway chain celebrated 30 years in business and witnessed the opening of its 11,000th restaurant.

As for the 17-year-old—he did complete his journey—he received a bachelor's degree in Psychology in 1971 and has become a successful businessman, to this day serving as President of one of the world's largest fast-food chains. Fred DeLuca believes in "valuing the experience of the journey rather than only visualizing its end." Today, he shares the journey and success with thousands of Subway franchisees and millions of Subway customers around the globe.

THE SUBWAY MURAL—HOW IT ALL BEGAN

Between 1965 and 1976, there was no unified decor concept among the Subway stores. It was in 1976 that Subway's President and cofounder Fred DeLuca decided to meet with a design company to develop a decorative theme that would help to establish a corporate image for the franchise chain.

Gordon Micunis Designs Inc. of Connecticut undertook the task of researching and developing a concept on which to base the store decoration. Gordon Micunis Designs proposed a "Decor Package," which used the Subway mural, depicting the construction of the New York Subway System, as a basis for the package.

The research that went into the creation of the original mural was quite extensive. Mr. Micunis felt that most people's thoughts of a subway were immediately associated with the New York City subway transit system. The "submarine sandwiches" had their origin in the Northeast, where people are familiar with the term *Sub;* however, people in the rest of the country might not identify the word with the sandwich.

Hence, it was determined that much of the artwork would relate to the building and creation of that subway system. Micunis derived most of the information from old newspaper clippings and photographs. The mural explores the complex engineering that went into the planning and construction of the system, the social comments, timetable of events, and the remarkable convenience afforded the subway traveler.

After months of compiling information, Micunis produced and installed the prototype Subway mural in the Ithaca Mall Subway store, located in Ithaca, New York. It was a huge success.

In 1992, Subway Art Director Ruth Woyciesjes introduced the Cityscape Mural as an alternative decor package. Woyciesjes incorporated a New York City landmark building into a collage of famous structures. The idea was to show the various historic buildings that were in existence during the 1920s and how the subway system connected people with the city above.

Today customers throughout the world are amazed and enlightened by the amount of information contained in both the original Subway mural designs.

(a)

(b)

An exterior view of one of the several types of SUBWAY restaurants (courtesy SUBWAY Restaurants)

INTERESTING FACTS

- Subway restaurants have been featured in many major motion pictures including: *The Coneheads, Lethal Weapon, Ace Ventura— When Nature Calls, The Beverly Hillbillies, Happy Gilmore,* and *Ransom.*
- Since 1994, Subway franchises have been voted America's favorite sandwich chain according to a poll conducted by *Restaurants and Institutions* magazine.
- Subway Sandwiches and Salads is the world's largest submarine sandwich chain and the second largest restaurant chain with more than 13,000 Subway restaurants in 62 countries.
- Subway restaurants serve a Chocolate Brazil Nut cookie, made with nuts harvested from the Peruvian rainforest. This environmental initiative helps to preserve the rainforest by providing more than 250 families with jobs that are not destructive to the rainforest.
- One can enjoy the greatest taste of Subway sandwiches while exploring the wonders of the Museum of Natural History in New Mexico or at Clyde Peelings Reptileland in Pennsylvania.
- The Subway chain has been ranked the number-one franchise opportunity by *Entrepreneur Magazine* for eight of the past ten years.
- The Subway menu offers seven sandwiches with 6 grams of fat or less.
- The Subway Read-A-Book program is one of the many ways that Subway restaurants work to promote education.
- Last year Subway customers consumed more than 60 million pounds of cold cuts.
- There are nearly two million different sandwich combinations available on the Subway menu.

CHAPTER 7

Standard Franchisor Services

Franchisors provide a variety of services to franchisees as a part of the franchise agreement. These services are the basic components of franchising and therefore need careful assessment by any interested franchisee. An examination of the franchise disclosure document reveals that a listing of services constitutes a substantial part of it. Franchisees pay a substantial amount for these services in the form of franchise, royalty, and advertising fees.

Not all restaurant franchisors provide all the services discussed in this chapter. Most restaurant franchises are business-format franchises providing a variable array of services. The manner in which they are performed delineates the efficiency of the franchise system. Providing services per the franchise agreement helps build strong franchisee-franchisor relationships. Conversely, if services are not efficiently provided, it leads to friction between the franchisor and the franchisees. The profitability of a franchise system depends on both the franchisees and the franchisor, and so these services assume considerable importance.

Franchisor services cover nearly the gamut of the restaurant business, including one-time services such as site and building development and on going services such as training, purchasing, marketing, and product development. To facilitate these services, many franchisors maintain regional offices in territories where their restaurants are located. Field staff personnel are employed to provide various functions of the services. Standard services provided by a franchisor are listed below and described in the following paragraphs.

- Site selection counseling and assistance
- Assistance in building and equipment
- Training at different levels for franchisees
- Pre-opening and restaurant opening assistance
- Continuing counseling in the operation of restaurant
- Providing operational manuals
- Know-how pertaining to the menu, ingredients/formula, and method of preparation
- Communication link between franchisor and franchisee
- Assistance in marketing, advertising, and promotion
- Permission to use trademark, service marks, and signage
- Franchise development and support
- Product development
- Purchasing and specifications
- Materials development
- Maintenance and inspection of standards and controls
- Operational support of field services
- Counseling on legal matters
- Financial assistance in maintaining and reporting transactions, accounting, and cost analysis
- Research and development
- Facilitating community activities and special events

SITE SELECTION AND RESTAURANT DESIGN

The first step in building a restaurant is its location. Site selection is important because the success or failure of a restaurant depends to a great extent on its location. A careful assessment of restaurant sites by an experienced professional is necessary. Established franchisors have expert real estate and property development staff and provide their assistance to franchisees. A complete market feasibility study is undertaken that includes data on overall market, population demographics, traffic patterns, site size and cost, break even sales, and competition.

After the site selection, the design of the restaurant is the next important matter. Design services are provided by franchisors, as described in the franchise package and information brochures given to potential franchisees. For example, McDonald Corporation, in its brochure, states that "The Company applies the same analytic approach used in selecting its sites to designing its buildings. An experienced staff of architectural, construction, and engineering personnel ensures that the McDonald's restaurant facilities are among the most technologically advanced and efficient in the foodservice industry, and are available in a variety of sizes and design that suit specific market needs." Some franchisees have different types of sites to provide flexibility in site selection and to serve varied market areas. For example, Applebee's® International offers prototypes ranging from 161 seats in smaller markets to 250 seats in markets with more demand. Golden Corral® has prototypes ranging from 5,000 to 10,000 square feet. The factors considered by Hardee's in site selection are general location and neighborhood, traffic patterns, access, competition, visibility, traffic generators, convenience of location, and size of site. Churchs® chicken considers the following characteristics, among others, in evaluation the site proposed by the franchisee: demographic characteristics (such as the number of households in the neighborhood, average income and family size), traffic patterns, proximity to existing restaurants, and the size and condition of the proposed premises. The factors considered by KFC® in approving a site selected by the franchisees include general location and neighborhood, traffic patterns, parking facilities, size, ingress and egress, visibility, demographics, and competitors' locations.

The complete package, including site selection, building design, interior layout, and decor, is what makes franchising so attractive to potential franchisees. Exterior and interior design become the symbols of franchise identification and are marketing features. Efficient operation, in addition to visual appearance, is considered while planning a franchise restaurant. Because the design of the restaurant is duplicated at different places, factors such as climate, soil conditions, and water table are considered. Also taken into consideration is that many franchised restaurants are located on choice, expensive real estate, where efficient utilization of space becomes a necessity. Thus, the building and equipment package is custom designed and provides all detailed specifications. Restaurant design and function have already been tested, in the case of established franchisors, and, hence, the franchisee is getting a proven design and layout. Working with known and proven specifications permits makes it easier to apply for building codes, permits, and lease agreements. Experienced guidance is available through the intricacies of the construction process.

Construction may be for a new restaurant or for the conversion of an existing structure to meet franchise standards. It can be a freestanding unit, located in shopping mall outlet, a food court, or a multiconcept unit within an existing facility designed for other business such as gas station and convenience store. Many franchisors have several designs and interior decor from which the franchisee may select. Franchisors' standards for the location, development, and construction of restaurants are based on their overall national marketing plans. Many franchisors retain ownership of the restaurant facilities and lease them to a franchisee. Although the franchisee pays for all expenses, there are rigid specifications that must be adhered to. Thus, signage, lighting, seating, decor, and overall construction must meet the franchisor's specifications. The franchise agreement grants the necessary rights and authorizations to operate a franchise restaurant facility. The creativity and input of the franchisees are also solicited by franchisors at times.

Geographic Distribution of Franchises

As a matter of corporate policy, a franchisor determines the geographic distribution of franchises. Not all franchises operate in every region or every state. Regions for franchises are selected for a variety of reasons. One way of selecting is based on areas of dominant influence (ADIs) designed by the Arbitron Company, a research division of Control Data Corporation. ADIs are commonly used in planning advertising and promotion. The ADI approach divides the United States into dominant television market areas. Each county is assigned to an ADI based primarily on television viewership. ADIs are used by some franchisors in making their franchising decisions. Consumers' buying power indices are also used for deciding on the number of franchised units desired. Other data related to demographics may also be used in making decisions to establish restaurants. Once a target area is selected, franchisees are recruited and assigned by the franchisor. The next step then involves site and individual market feasibility studies.

Factors to Be Considered in Site Analysis

Factors to be considered in site analysis are shown in the checklist presented in Figure 7.1 and described in the following paragraphs.

> *Zoning.* Zoning is one of the most important considerations for commercial restaurants. Areas have specific zoning laws, which differ in definition and interpretation. It is essential to know exactly what the available zoning permits allow. Many aspects of a site are regulated by zoning laws, such as the

CHARACTERISTICS	COMMENTS
Date: _____ Proposed Site: _____	
Types of Foodservice: _____	
ZONING	
1. Current zoning	_____
2. Anticipated changes in zoning	_____
3. Height restrictions	_____
4. Parking restrictions	_____
5. Back and side yard restrictions	_____
6. Sign restrictions	_____
7. Other restrictions	_____
LOCATION. Driving time and/or distances from:	
1. Residential areas	_____
2. Office complexes	_____
3. Business districts	_____
4. Educational facilities	_____
5. Major market(s)	_____
6. Sports and recreational activities	_____
7. Historical sites and attractions	_____
8. Interstate highways	_____
9. Industrial centers	_____
10. Shopping centers	_____
AREA	
1. Type of population	_____
2. Future growth pattern	_____
3. Type of businesses	_____
4. Development of nearby areas	_____
5. Target population(s)	_____
6. Labor outlook	_____
7. Planned development(s)	_____
PHYSICAL CHARACTERISTICS	
1. Top soil	_____
2. Subsoil	_____
3. Water table	_____
4. Surface drainage	_____
5. Slopes	_____
6. Landscaping	_____
7. Elevation	_____
8. Distance from river, creek banks	_____
9. Other characteristics	_____
LAND MEASUREMENTS	
1. Length	_____
2. Width	_____
3. Total area	_____
4. Total usable area for: Building	_____
Parking	_____
Open space	_____

FIGURE 7.1 Market and site analysis form

CHARACTERISTICS	COMMENTS
VISIBILITY 1. Obstructions 2. Visibility from different directions 3. Visibility of signs 4. Visibility affected by location (e.g., within a mall or a tall structure) 5. Improvements needed for visibility	
TRAFFIC PATTERNS AND REGULATIONS 1. Traffic counts on site street 2. Traffic counts on closest main street 3. Peak traffic timings 4. Type of traffic 5. Distance to nearest highway 6. Traffic regulations: one-way, stop-signs, no-turns, speed limits 7. Parking regulations 8. Public transportation	
SERVICES 1. Police 2. Fire 3. Trash 4. Garbage 5. Security	
STREETS 1. Width 2. Pavements 3. Curbs 4. Sidewalks 5. Lighting 6. Grades 7. Hazards 8. Overall conditions	
UTILITIES 1. Water 2. Sanitary sewer 3. Storm sewer 4. Electricity 5. Gas 3. Steam	
COMPETITION 1. Number of food facilities 2. Type of facilities and menu 3. Service style 4. Number of seats 5. Average sales 3. Type of competition 3. Impact of competition	

FIGURE 7.1 *(Continued)*

height of the structure and backyard and sideyard requirements. From the restaurant point of view, zoning laws also control two other important aspects: parking and use of signs. Laws specify the minimum number of parking spaces required for a restaurant on the basis of the number of guests that can be served. Some areas have restrictions on the size, height, and type of signs that can be used, whether as trademark displays or as advertisement. Liquor permits are also sometimes based on zoning laws for the area.

Area Characteristics. The profitable functioning of a restaurant is dependent, to an extent, on the characteristics of the area. The type of location, such as interstate highways, educational campus, mall, or food court, provides preliminary information on the type of consumers that can be expected. One of the factors taken into consideration is the growth potential and growth pattern of an area. Many franchisors and franchisees have benefited from accurate prediction of the growth pattern of a location. Future development of industrial complexes, shopping centers, a main highway, resort areas, sites for entertainment facilities, and newly constructed subdivisions are all taken into consideration when choosing locations for franchise restaurants.

Physical Characteristics. Certain physical characteristics provide clues to the suitability of the site for restaurant construction. Low-lying areas with poor soil drainage may pose problems during possible flash flooding. Similarly, the depth of the water table dictates whether a basement can be used for storage or other purposes. Landscaping is also extremely important. Natural landscaping, trees, and lakes not only improve the aesthetic and commercial value of a restaurant but also allow for extra facilities, such as a play area for children.

Cost Consideration. The cost of the land and the cost of improvement are important considerations. Renovations and modifications may prove expensive, so costs are calculated carefully. It must be remembered that restaurants have special requirements, and not all existing buildings are suitable for restaurants without extensive alterations.

Utilities. Utilities play an essential role in any restaurant operation, and access to energy and the type of energy available are important. The location of major utilities such as electricity, gas, telephone, water, and steam must be considered. Access to storm sewers must also be noted. Once utilities are installed, recurring costs may be involved, so it is essential to check beforehand. Drainage and sanitation should be considered in light of local public health regulations.

Access. Access routes to restaurants are important, particularly in areas where severe weather occurs. The type and the condition of streets, curbs, gutters, and pavements must be studied. The types of transportation available (i.e., bus, train) also are significant, as they are necessary not only for consumers and employees but also for deliveries. In addition, it is necessary to consider street lighting and parking lights.

Position of Site. Location of the restaurant must be considered on the basis of driving distance as well as time to and from industrial, residential, recreational, sports, educational, and business centers. The proximity of the restaurant to these areas can give the planner some idea of the number of consumers that can be expected, so a careful assessment of the site characteristics is imperative.

Traffic Information. In addition to site characteristics, traffic flow patterns are important. Traffic counts are recorded and used in the final analyses of the site data. A survey of traffic patterns indicates the time and direction of traffic flow. In addition to the pattern, it is also necessary to measure the frequency patterns of traffic flow. One-way streets, speed limits, and availability of parking influence a patron's decision to visit a restaurant.

The types of transportation normally used by patrons, such as cars, buses, and trucks, must also be taken into consideration. Anticipated changes in the flow of traffic should be studied.

Availability of Services. Access to services is a factor often ignored in the analyses of data. One of the most important services for a restaurant operation is trash and garbage pickup. Frequent waste removal is highly desirable. In addition, such services or service facilities as police, security, fire stations, fire hydrants, and sprinklers must be checked.

Visibility. A restaurant can be greatly enhanced by good visibility, which is extremely important on highways and in remote areas. High-placed and lighted signs on highways are major attractions. Some states permit food facilities to advertise on road sites at appropriate intervals. Sometimes, wooded or shaded areas can be arranged to provide landscaping in addition to good exposure for the restaurant. Any obstruction of signs or visibility needs to be checked. Signs—their location, type, spacing and size—are all-important considerations.

Competition. It is obvious that a restaurant operation, to be successful, must consider its actual and potential competition. Major competitors must be considered in the light of their number, number of seats, turnover rate, types of menus offered, average check, and annual sales. A restaurant may prove unsuccessful if the potential competition is not adequately evaluated.

Market. Information pertaining to consumers must be collected, including data about age, sex, occupation, income, food preferences, access, transportation facilities, and potential for future growth and development.

Type of Restaurant and Service. The type of restaurant and service provided must be taken into consideration. For example, a pizza restaurant, coffee shop, drive-in, counter service, and hamburger stand all need different types of operational design.

Restaurant Design: Important Factors to Be Considered

The following points must be considered in designing a restaurant:

- The exterior should be inviting and present the signage of the franchise.
- Exterior portions and entry to the restaurant must be neat, clean, and, possibly, artistically designed.
- Drive-in facilities must be well planned and should not lead to traffic congestion or jeopardize the safety of the customers.
- The receiving area should be located away from and preferably out of sight of the main guest entrance or dining area.
- Storage areas should be neatly arranged and as close to the kitchen area as possible.
- All equipment should be arranged in a sequential order and should facilitate efficient food flow.
- Space should be allocated based on the importance and priority of each function.
- Safety of the employees must be considered in placing and operating equipment.
- The sanitation area and facility must be within reach of all employees.
- Primary and secondary colors must be effectively combined to provide a lighted, relaxing atmosphere within the restaurant.
- Comfort of the patrons in relation to the inside temperature, sound, seating, and odor level must be carefully planned.
- The design and decor must be well coordinated. It is also desirable to have a theme around which the restaurant is planned.

TRAINING

The success of any franchise system depends on the training program provided by the franchisor. As described in earlier chapters, the success of a franchising system is based on the uniformity of its products and services. This uniformity can only be achieved by an effective training program.

This training program varies from franchisor to franchisor. An excerpt pertaining to training in a McDonald's brochure is cited here as an example:

> Should both McDonald's and the applicant agree to proceed, the formal training begins. This program, which typically takes in excess of 24 months, again normally takes place on a part-time basis and requires approximately 20 hours per week. The program involves several phases, most of which take place in a McDonald's restaurant convenient to the candidate's home. Interspersed with the in-store training are four formal classroom sessions of one- to two- week duration—one of which takes place at the Company's Hamburger University. There are also seminars, conferences, one-on-one sessions with corporate personnel and sophisticated audiovisual training.

There are several advantages of having a well-planned and organized training program, some of which are:

- The franchisor is able to explain the concept, philosophy, and operations of a franchise to the franchisee.
- The franchisee gets hands-on experience in restaurant operation and management.
- Provides franchisee the opportunity to assess if this is the kind of business that he or she would like to invest in or get involved with.
- Indicates the capability, or lack thereof, of the franchisee to successfully operate the franchisor's business.
- By anticipating questions, reduces the number of inquiries when the restaurant is in operation.
- Motivates franchisees to perform at their best level once they understand all aspects of the franchise business.

- Increases satisfaction of the franchisee as well as the employees working for the franchise unit.
- Reduces complaints from consumers and employees.
- Helps maintain quality of products and services based on the standards set by the franchisor.
- Promotes adherence to sanitation standards in all functional areas of the business.
- Reduces breakage and spoilage within the franchise unit operation.
- Reduces number of accidents.
- Creates an identity for the franchisee within the franchise system and fosters development of franchise loyalty.
- Improves operational skills of the franchisee.
- Establishes the franchisor and the franchisee as a team rather than two separate partners.
- Opens a dialogue between the franchisor and the franchisee.

In summary, a training program offers many advantages. Training helps in comprehension of the franchise concept, in understanding its operation, and in promulgation of standards and uniform operating procedures.

Types of Training Programs

Normally training offered by franchisors is of several types, including preopening, opening, and continuing training. These programs may be structured for potential franchisees/owners, for restaurant operators, or for crew members. The contents of the training deal with developing technical and management skills. In addition, orientation to the company philosophy, operational parameters, and human resource skills are included in the course contents.

Potential franchisee training or preopening training program is primarily for the purpose of evaluating the potential of the franchisee to successfully operate a franchise restaurant. This training may be provided at the corporate headquarters training center, at a regional training center, at a field representative office, and/or at a franchise restaurant located anywhere in the nation that may be closer to the potential franchisee's home. Some franchisors have well-established training centers. Although franchisees pay for transportation and lodging, most franchisors provide training and training materials. Irrespective of the cost, the worthiness of the training program both for the franchisees and the franchisors cannot be overestimated. The content of a training program usually consists of the corporate philosophy and organization, a description of restaurant operations, and hands-on experience related to products and services offered at the franchise. A training manual that describes the training in detail is used for this program. The training manual becomes a continuing source of information and can be used as a reference by the franchisee.

Final selection of franchisees is based on successful completion of this training program. The duration of training varies among franchisors. KFC restaurants require an initial training program of approximately seven weeks in duration, with classes averaging about eight to ten hours per day. During the initial training course, approximately five days are spent in a classroom and thirty days in a KFC outlet. Training center instructors conduct classroom segments and field training instructors coordinate training in KFC restaurants. In addition to the initial training, KFC offers assistance in the areas of customer service, general outlet management, quality control, employee training, new product preparation, and equipment maintenance. The new franchisee orientation program before opening a restaurant offered by Churchs Chicken consists of one forty-hour week of workshops and seminars conducted at a training facility designated by the company.

The number of persons required by the franchisors to attend the training programs ranges from two to four. Franchisors hold the right to designate other persons than the franchisee to undergo the training for specific activities. Managers, supervisors, and their replacements are encouraged complete training. Instructors, training materials, and other supplies are generally provided by the franchisors. Travel and living expenses incurred in connection with training are borne by the franchisees.

Restaurant Operator Training is designed for those who will operate the restaurant, who may or may not be franchisees themselves. In this training, the emphasis is on the operational aspects of the restaurant franchise. Trainees are exposed to the management of operations, human resources, cost accounting, basic equipment maintenance and operation, restaurant closing-out, and recordkeeping. A small team of operators, rather than a single person, is recommended by some franchisors to attend this training.

An *initial opening training program* is designed to provide assistance during the opening period of the restaurant. A franchisor's representative is on hand at the new restaurant for the first few days to assist and train in all aspects of the initial store opening. This on-site training assistance is valuable and, because the opening period is critical for the franchisee, it is usually welcomed. Unexpected problems and complexities can be handled at this stage of the business by customized training. Some representatives also train the staff of the restaurant during this time.

A *crew member training program* may be offered at the franchise unit and may include classroom courses, videotapes, films, interactive video programs, and programmed materials. The emphasis of this training is on functional areas of the restaurant, such as service and sanitation. Training modules are also developed to teach crew members aspects of daily restaurant operation. An example of such a program is "Training by the Slice," offered by Godfather's Pizza.

Ongoing training programs are offered by some franchisors. This training is provided on a regular basis and may consist of classes, seminars, conferences, and workshops. Training can be provided on site, at regional field offices, or at corporate training centers. Ongoing training helps keep franchisees abreast of the changes and development taking place in any aspect of the franchise business. The requirements and contents of ongoing training are very similar to the pre-opening training. Advanced-level training is also provided for upper management employees. Specific courses pertaining to regulations originating with Occupational Safety and Health Act (OSHA), Americans with Disabilities Act (ADA), etc., are also offered.

Most franchisors consider training the top priority among the services provided to franchisees. In planning training programs, established principles of teaching and learning should be employed. Franchisees are adults who may have different educational levels; this must be considered in planning training. Periodic evaluation of training programs provides valuable feedback for improvement and modification.

Marketing Support

Restaurant franchises rely heavily on advertising and promotion, and a significant portion of the budget is always earmarked for advertising. Franchisees pay an average of about 4 percent of gross sales as advertising fees. Franchisors maintain well-qualified marketing staff to help in all aspects of marketing, such as advertising, promotion, and public relations. Efforts are focused toward marketing at international, national, regional, and local levels. It is always advisable, particularly at the local level, that franchisees be consulted about advertising materials to be used for television and radio commercials, print ads, and promotional materials. Pooled funds from franchisees yield a combined buying power that strengthens the franchisors' efforts in effective marketing.

Advertising focuses on the entire business format, whereas promotion is geared toward a specific product or menu mix. Advertising can be done as mass advertising, in which the message is communicated to the target customer using mass media such as television, radio, billboard, and telephone books. Mass advertising is expensive for individual franchisees, which makes the franchisor's pooled resources a crucial resource. Advertising on a special occasion, such as during the televised Super Bowl, is very expensive but reaches a large target audience. Mass advertising reaches a large audience in a broad area and helps build the franchisor's image.

The most common media used for mass advertising are television and radio. Print advertising is also widely employed by franchisors and uses a wide variety of media. Promotion of specific items is undertaken by organizing sales promotions during special occasions and seasons. Promotion can be done nationally but is most common at the regional level. These promotions, while featuring specific items, such as special sandwiches, help generate repeat business and build consumer loyalty. Some franchisors provide merchandise planners and source guides to assist franchisees in conducting their own related promotional sales. Coupons are most commonly used for specific promotions.

Target marketing programs are also used by franchisors for specific population segments. These are customized programs designed to attract a special category of consumer. As an example, many quick-service franchises have created cartoon characters to attract children. Happy Meals™ (McDonalds) is an important promotion geared toward a specific target market. Point-of-sale promotions are also conducted; these are in-store promotions to sell particular products to walk-in or drive-in customers. Point-of-sale promotions are done via banners, menu clips, displays, and handouts.

In summary, franchisors provide marketing staff that helps in marketing programs and conducting market analysis and research. These staff help both the franchisor and the franchisee to forecast potential changes and marketing opportunities and to react quickly and accurately to marketplace changes. Their major function is to continually collect data on consumer attitudes and usage patterns and to analyze those data in a useful and meaningful fashion. Some franchisors have field marketing managers or executives who can be approached by the franchisees to plan their own advertising and promotional activities. This advice can be very useful for franchisees, particularly during the restaurant opening period.

For the purpose of advertising, arrangements are made whereby national cooperatives are responsible for producing, disseminating, and placing national advertising via print, radio, television, magazines, newspapers, billboards, and other media.

All marketing efforts must be designed to meet the organizational objectives of the franchise system and should be based on consumer needs. Franchisees should understand and select the best marketing strategies and choices for their particular operation. Some franchisors maintain marketing advisory committees that consider and make decisions on matters relating to marketing strategies. Marketing is an important aspect of franchisor assistance and leads to profitability of the entire franchise system.

MATERIALS MANAGEMENT

Materials play a key role in the restaurant business. Materials include ingredients, supplies, and equipment. Because quality and consistency are essential for the success of a franchise, uniform ingredients and equipment are indispensable. Franchisors may provide different options for this purpose, some of which are listed below.

Some franchisors make it mandatory that all ingredients, supplies, and equipment be purchased from them. This not only helps in maintaining consistency in operations and facilities layout but also results in savings due to the combined purchasing power of the franchisor. This is an advantage to franchisees *if* properly managed and administered. Certain proprietary formulas for ingredient mixes, make it mandatory for franchisors to sell those products to franchisees. For example, some franchisors' doughnut mix or frozen yogurt mix are based on proprietary formulas that cannot be revealed to franchisees.

Franchisors require that *some* ingredients, supplies, and equipment be purchased from the franchisors. This requirement may apply to equipment only or to certain ingredients. The rest of the items can be purchased per the specification of the franchisor. For example, bread and baked items may be purchased from local bakeries that can meet the franchisor's specifications. The consistency of products and services is controlled to a degree by this arrangement. Franchisors maintain quality assurance and materials management through a program by which only ingredients or products from approved suppliers or distributors are allowed in the franchise system. Regional field staff inspect franchises to see that only approved products are used.

Some franchisors have centralized distribution centers that supply approved specified products to franchisees. This type of centralized distribution eliminates the need for franchisees to deal with numerous suppliers, to check for product consistency, and to follow cumbersome billing procedures, allowing managers to devote their time and effort to other management functions. Combined purchasing helps a franchisee to purchase these items at competitive prices. These centers are placed in geographically convenient locations.

Certain restaurant franchisors maintain equipment engineering staff who design, develop, and test equipment that is most appropriate for the menu offered. This service may result in nonconventional equipment that is more energy-efficient, cost-effective, and suitable for the menu.

OPERATIONAL SUPPORT AND FIELD SERVICES

Operational support is needed by franchisees for occasional questions and problems. Many franchisors have regional staff trained to assist franchisees in every aspect of restaurant operation to provide this support. Moreover, they are familiar with common problems because they regularly handle queries from different franchisees.

Normally these representatives can be reached by toll-free numbers nearly twenty-four hours a day. These services help maintain optimum quality of products and services. Inspections and follow-up reports are provided to improve unit performance.

A strong communication link is established between franchisor and franchisee by these field services. Any new product or service introduced by the franchisor is integrated into the restaurant operations with field service assistance. Exchange of information and ideas to and from franchisees can also take place through these regional offices.

Financial Control Assistance

Some franchisors assist in setting controls pertaining to cost and inventory levels. They assist franchisees or their accountants in setting up accounting systems and preparing financial statements. Assistance is also provided in financial reporting and analysis of financial statements. Franchisors may also provide assistance in computerized on-line data maintaining systems. These systems help in controlling food and labor costs and maintaining optimum inventory level, which is particularly helpful if the franchisee is a multi-unit operation. A comparative assessment of the financial position of a restaurant can also be obtained by using figures from other units.

Assistance is also provided by franchisors in assessing the financial performance of a franchise. A program for maximizing profit potential by controls may also be designed for use by the franchisees.

Research and Development

Restaurant franchises face tough competition and new products are constantly being tested and introduced into the market. This competition requires continuing research and development (R&D) activity. All major restaurant franchisors maintain efficient R&D services. New menu items are tested almost daily for popularity and applicability. Some of the currently most popular menu items were introduced as a result of R&D efforts. Research and development units are normally located at corporate headquarters or where access to company-owned units is convenient. Industry trends, consumer attitudes and preferences, and the franchise concept are all taken into account by the R&D department.

A function of this department also is to test ingredients and products on a random basis for quality. The R&D staff also assist in solving product and ingredient problems as they occur in individual units. Equipment testing and development is also conducted.

Thus, new products, ingredients, equipment, and processes are always under research for further development, resulting in convenience to consumers and employees, new items on the menu, and improved quality of products and services provided by the franchise. Research and development helps franchisees by resulting in better service to the consumer and increased profitability.

COMMUNITY INVOLVEMENT SUPPORT

Franchisors encourage and at times support civic and charitable activities. A positive image and a sense of belonging are developed by the involvement of the franchise system in community activities. Involvement is one way franchisees can return something to the communities in which their businesses are located. Many franchisors support college and school sports teams and events. Scholarships and donations are provided by the franchisors, which helps build the image of the franchisee in the community. Support in controlling environmental problems and furthering social improvements is particularly appreciated by the public.

FRANCHISOR'S OBLIGATION

Franchisors are obligated to provide services mentioned in this chapter. These obligations are listed and described in the franchise offering circular. The assistance, supervision, and other services provided by franchisors to franchisees are divided as follows:

Before the restaurant opens the franchisor generally provides the following services to the franchisee:
- Furnishes a description of the general area in which the franchisee may establish an outlet.
- Notifies whether or not the site proposed by the franchisee and related plans are acceptable. Each site proposed by the franchisee is evaluated and written notices are given pertaining to its acceptance or nonacceptance.
- Furnishes standards and specifications regarding building type, access requirements, furnishings, and equipment. These plans are for use in the construction and renovation of the restaurant and cannot be modified or deviate from the original plans.
- Furnishes a list of approved suppliers for equipment and construction materials.
- Provides training and operating advice. This initial training program is offered to franchisees and some of their designated employees.
- Loans a copy of the operating manual (Some franchisors call this by different names; for example, KFC refers to its operating manual as (the *standards library*). This manual is considered confidential because it contains functional information about the restaurants, company policies, and operating procedures.
- Provides a representative to be present at the opening of the first restaurant.

During the ongoing operation of the restaurant the franchisor agrees to:

- Offer continuing training programs deemed necessary by the franchisors.
- Provide continuing services deemed necessary for the business. This includes refinement of products and equipment, informing the franchisee about quality control methods, and research and development.
- Work to maintain sources of supply for items incorporated in the franchisor's trade secrets, which are essential for operating a restaurant outlet.
- Make available continuing advisory assistance in the operation of the restaurant.
- Continue efforts to maintain high and uniform standards of quality, cleanliness, appearance, and services at all restaurants.
- Conduct periodic inspections of the premises of the restaurant and periodic evaluations of the products used and sold at the restaurant.
- Assist in advertising and marketing.
- Teach how to manage controllables, such as food cost, labor, and inventory.
- Provide counsel on how to best increase sales for the restaurant.
- Assist in the purchase of commodities and equipment.

The above lists are comprehensive but do not imply that each franchisor provides all of the services. For example, Hardee's does not select the location for the restaurant, but it does provide site selection assistance upon the franchisee's request. This assistance consists of overall evaluation of alternative sites and help in preparation of materials to be submitted to Hardee's as part of the agreement.

Restaurant Study 11

Made in Japan Teriyaki Experience

A Brief History

In 1983, The Donato Group of Companies conceived an idea for a Japanese fast-food restaurant. The concept would provide dishes similar to that of expensive Japanese sit-down dinner houses but at fast-food portions and pricing. After three years of extensive research and fine-tuning, Made in Japan Teriyaki Experience was born. The first restaurant opened in August of 1986. In 1987, with the success of the first two corporately owned stores, a franchise expansion program was developed.

The focus of Made in Japan has always been on fresh vegetables, Yakisoba noodles, Japanese sticky rice, a wide selection of grade-A, quality meats and seafood, and sushi. By preparing meals directly in front of the customer, they are able to keep their clients entertained at the same time as reinforcing the freshness of everything they serve. Finally, the "uniqueness" of their very special teriyaki sauce has turned it into a signature product which many competitors have tried, unsuccessfully, to duplicate.

Today, Made in Japan is the largest Japanese fast-food restaurant chain located in food courts in Canada. Now, their focus is not only food courts. Made in Japan is aggressively targeting other high-traffic locations, such as airports, universities, office buildings, busy city streets, as well as train and bus terminals across Canada and around the world. Their goal is to build brand recognition around the world through the careful selection of dedicated Master Franchisees and Area Developers.

Standard Franchisor Services 205

(a)

(b)

An interior view of Made in Japan Teriyaki Experience

Restaurant Study 12

Burger King Corporation

A leader in today's fast-food industry, with locations in all 50 states and 56 international countries and territories around the world, Burger King Corporation was founded in 1954 in Miami, Florida, by James McLamore and David Edgerton.

McLamore and Edgerton, both of whom had extensive experience in the restaurant business before starting their joint venture, believed in the simple concept of providing the customer with reasonably priced quality food, served quickly, in attractive, clean surroundings.

Since its Florida beginnings more than 43 years ago, when a Burger King hamburger cost 18 cents and a Whopper sandwich cost 37 cents, Burger King Corporation has established restaurants around the world—from Australia to Venezuela. By 1967, when the Company was acquired by the Minneapolis-based Pillsbury Company, 8,000 employees were working in 274 different restaurant locations. Today, there are more than 300,000 Burger King employees in more than 9,700 locations worldwide.

The success and size of Burger King are the result of a tradition of leadership within the fast-food industry in such areas as product development, restaurant operation, decor, service, and advertising.

Just as the Whopper sandwich was an immediate hit when it was introduced in 1957, each of the Company's products provide the quality and convenience sought by today's consumers. The BK Broiler, a grilled chicken sandwich introduced in 1990, sells up to a million a day. Still, the Whopper sandwich, one of the best known hamburger sandwiches in the world, remains a perennial favorite. More than 1.6 billion Whopper sandwiches are sold annually.

One of the factors that has helped to increase the Company's expansion and growth has been the sale of restaurant franchises. By 1961, McLamore and Edgerton had acquired national franchise rights to the Company, which was then operating 45 restaurants throughout Florida and the Southeast.

Restaurant decor has traditionally been important in creating memorable images for Burger King consumers. Burger King was the first fast-food chain to introduce dining rooms, allowing customers a chance to eat inside. Drive-thru service, designed to satisfy customers "on-the-go," was introduced in 1975, and now accounts for approximately 50 percent of Burger King business. "Take-out" represents another 15 percent of off-premise dining.

Burger King Corporation has always taken great care in the design and construction of its restaurants so they will be attractive features of their communities. The "Double Drive-Thru," "Kiosk", "In-Line," and "Co-branded" restaurants are among the new concepts that will propel Burger King forward to meet future challenges of the 1990s.

Burger King Corporation's advertising campaigns have also contributed to the Company's success. The Company's first television ad ran on Miami's only VHF station in 1958. One year after The Pillsbury Company acquired Burger King as a subsidiary in 1968, the Company's first major promotion, "The Bigger the Burger the Better the Burger," debuted. In 1974, the memorable "Have It Your Way" campaign was created. Advertising campaigns in the last ten years have included "America Loves Burgers and We're America's Burger King"; "Best Darn Burger"; "Make It Special, Make It Burger King"; "Aren't You Hungry"; "Battle of the Burgers"; "Herb"; "Burger King Town"; "Best Food for Fast Times"; "We Do It Like You'd Do It"; "Sometimes You've Gotta Break the Rules"; "Your Way, Right Away"; "Get Your Burger's Worth"; and most recently, "It just tastes better."

In 1985, average restaurant sales passed the $1 million mark and a European training center opened in London to service overseas Company and franchise employees.

Company management, which had been housed under one roof in the Kendall area of Miami since 1970, moved in 1988 to an 114-acre site in southern Miami-Dade County, Florida. The World Headquarters consists of three buildings, with 290,000 square feet of office space. Nearly one-half of the total site was preserved in its natural state.

Perhaps the most impressive trait that Burger King Corporation possesses is its desire to continually enhance the product line and brand image. And in today's brand-conscious society, this can only translate into continued success.

Burger King Corporation and its franchisees operate more than 9,700 restaurants in all 50 states and 56 countries around the world. In fiscal year 1997, Burger King had systemwide sales of $9.8 billion. Burger King Corporation is a subsidiary of Diageo plc, one of the world's leading branded consumer products businesses. Diageo's portfolio of international food and drinks brands include Guinness, Pillsbury, Green Giant, Häagen-Dazs, Old El Paso, Progresso, Burger King, Smirnoff Vodka, Bailey's Original Irish Creme, and J&B Rare Scotch Whisky.

Fast Facts for the 90's

- There are more than 9,600 Burger King restaurants in 56 countries and international territories worldwide.
- Over 4.6 million Whopper sandwiches are sold each day at Burger King restaurants around the world—more than 1.6 billion each year.
- Annually, 693 million hamburgers are sold at Burger King restaurants worldwide Daily, that's 20 semitruckloads of beef.
- Burger King Kids Club was launched nationally in January 1990. More than 1 million kids signed up in the first two months. There are currently 5 million members.
- Burger King Corporation's promotion of the Disney film, *Toy Story*, was an overwhelming success. More than 43 million premiums and 15 million trading cards were distributed in 5 ½ weeks.
- Each year, 1.7 billion units of french fries are sold at Burger King restaurants across the United States.
- Burger King uses more than 700 million pounds of fries annually in the United States.

Burger King Corporation—Historical Fact Sheet

1954   James W. McLamore and David Edgerton cofound Burger King of Miami, Inc., which becomes Burger King Corporation in 1972.
       McLamore and Edgerton's first restaurant, located at 3090 NW 36th Street, Miami, Florida, sells 18 cent broiled hamburgers and 18 cent milkshakes. Burger King offers 12 oz. regular and 16 oz. large sodas.
1957   Whopper sandwich introduced . . . appears on the menu for 37 cents.
1958   Miami-based Hume, Smith, and Mickelberry hired as first major advertising agency. They developed "Burger King, Home of the Whopper" campaign.
1961   McLamore and Edgerton acquire national and international franchising rights.
1963   Burger King goes international . . . two restaurants open in Puerto Rico.
1967   The Pillsbury Company acquires Burger King Corporation as a subsidiary for $18 million.
       274 restaurants in operation with a total of 8,000 employees.

1968	Ad agency Batten, Burton, Durstine and Osborne (BBDO) hired to develop first major promotion "The Bigger The Burger The Better The Burger."
1974	"Have It Your Way" campaign created by BBDO.
1975	First European Burger King restaurant opens in Madrid, Spain. Drive-thru service was introduced.
1976	JWT named as one of Burger King Corporation's advertising agencies.
1977	2,000th Burger King restaurant opens in Hawaii, putting locations in all 50 states.
1982	Burger King introduces Bacon Double Cheeseburger. Project BOB (Battle of the Burgers) advertising campaign. Late-night Drive-Thru introduced.
1983	Salad Bar debuts nationally. First on-campus Burger King restaurant opens at Northeastern University, Boston, Massachusetts. First mobile restaurant unit, the "Burger Bus" opened by Ohio franchisee. UniWorld Group, Inc., named as one of Burger King Corporation's advertising agencies.
1985	Breakfast debuts nationally with the Croissan'wich as the key product. Self-serve drinks introduced. Crew Educational Assistance Program (CEAP) established, providing crew members with $2,000 for postsecondary education opportunities. Ground breaking for new World Headquarters site on 114 acres (building sits on 40 acres) in Miami. Scheduled completion early 1988.
1986	A record 546 new restaurants open worldwide. 4,743 restaurants in operation . . . $4.5 billion in system-wide sales. 402 international restaurants in 25 countries. Chicken Tenders debut. New breakfast product (French Toast Sticks) introduced nationally.
1987	New breakfast product (Bagel Sandwich) introduced to menu. NW Ayer named as one of Burger King Corporation's advertising agencies.
1988	New World Headquarters opens on 114-acre site (building sits on 40 acres) in southern Miami-Dade County, Florida.

Grand Metropolitan PLC acquires The Pillsbury Company and its subsidiaries, including Burger King Corporation, for $5.79 billion.

Burger King begins regional roll-out of Chicken International sandwiches.

1989 Hispanic advertising agency Sosa, Bromley, Aguilar & Associates joins D'Arcy Masius Benton & Bowles, Saatchi & Saatchi Advertising, and UniWorld Group, Inc., as Burger King Corporation's advertising and marketing team.

Burger King launches ten Burger King Academies across the United States; BK Academies are alternative schools for children at risk of dropping out.

"Sometimes You've Gotta Break the Rules" ad campaign is introduced.

Burger King continues European expansion with conversion of nearly 100 Wimpy counter service restaurants in the United Kingdom to the Burger King brand. Grand Metropolitan acquired Wimpy restaurants in the August acquisition of UB Restaurants. A total of 200 restaurants were converted by summer 1990.

1990 Burger King Kids Club program launched nationally. One million kids register in first two months.

Burger King introduces BK Broiler, a flame-broiled chicken sandwich.

Burger King switches to 100 percent vegetable oil for frying french fries.

Burger King opens in Dresden, East Germany.

1991 "Your Way, Right Away" campaign is launched.

International expansion continues with Burger King restaurants opening in Budapest, Hungary, and Mexico.

1992 Burger King Corporation teams with Disney for the first of nine promotional theatrical releases tie-ins with the movie giant featuring the animated films *Beauty and the Beast* and *Pinocchio.*

Burger King opens its first restaurant in Warsaw, Poland.

1993 Burger King opens its first restaurant in Saudi Arabia.

Burger King Corporation becomes the largest single circulation publisher of children's magazines with the release of three new magazines, distributed to more than three million members of its Kids Club program.

	First international Burger King Academy opens in London, England.
	Burger King launches the Everyday Value Menu.
1994	Ammirati & Puris/Lintas named as Burger King Corporation's general market advertising agency.
	"Get Your Burger's Worth" campaign is launched.
	Burger King Corporation teams with Disney for the blockbuster hit *The Lion King*.
	The BK Broiler, the BK Big Fish, and the hamburger are increased by more than 50 percent adding real value to the customer.
	Burger King Corporation opens its first restaurant in Israel, Oman, Dominican Republic, El Salvador, Peru, and New Zealand.
1995	Burger King opens a record 657 new restaurants worldwide.
	Burger King enters Paraguay and Turkey.
1996	Burger King Corporation acquires 57 restaurants from Davgar Inc., making it the single largest purchase of company restaurants in Burger King history.
	Burger King employs 300,000 employees systemwide.
	Urban City Foods announces plans to develop 125 restaurants in inner cities over the next five years.
	Burger King Corporation teams with Disney for the movie favorite *The Hunchback of Notre Dame*.
	Burger King announces promotional partnership with Universal Studios for *The Lost World,* the sequel to *Jurassic Park*.
1997	President Bill Clinton selects Burger King Corporation to work with the White House in identifying ways to transition current welfare recipients into the work force. Burger King establishes its welfare-to-work coalition.
	Burger King launches its second signature hamburger sandwich, the Big King.
	Burger King launches its new hotter, crispier, and tastier french fries with a $70 million marketing campaign, making it the largest product launch in company history.
	Grand Metropolitan, Burger King Corporation's parent company, merges with Guinness to create a new company called Diageo, plc.

1998   Burger King launches Cini-minis in the United States. The product, which features the heart of the cinnamon roll, was developed exclusively for Burger King.
Burger King opens its 50th restaurant in Turkey.

(a)

(b)

An exterior view of a typical Burger King restaurant (courtesy Burger King Corporation)

CHAPTER 8

Financial Aspects of Franchising

Franchising requires investment from all parties, first from franchisors and later from franchisees. In addition to one-time fees, franchisees are required to pay different fees as outlined in the franchise agreement. It is imperative for a prospective franchisee to fully understand his or her financial obligations. In the case of restaurant franchises, the franchisee incurs costs related to purchasing the site of the restaurant, building or leasing the property, equipment purchase, inventory and supplies, legal expenses, pre-opening expenses, and a score of other expenses, and yet must have working capital on hand to start the business. Based on the type of franchise, the initial investment may range from a few hundred to a million dollars. In order to understand the various components of franchise investment, it is essential to understand the various fees and costs involved. The fees and costs commonly required by franchisors are explained in this chapter.

FRANCHISE FEE

This initial fee, also referred to as the *license fee,* is charged by the franchisor and can range from few hundred dollars to several thousand. Normally, for a well-established franchise restaurant the franchise fee averages $25,000. This fee is primarily to compensate the franchisor for the use of its trademark as well as to defray costs incurred in setting up a system to sell and market franchises. Although the franchise fee is a one-time fee, it is only for the period for which the agreement is signed. Normally it is payable at the time of signing the franchise agreement and is nonrefundable. In certain instances, when it is not possible to get zoning permits or there are building restrictions, the franchise fee is refunded. This fee also varies with the type of franchise. It is relatively higher for freestanding restaurants than for newer concepts, such as double-drive-throughs and kiosks. Also, based on the duration of the contract, the initial franchise fee can be prorated. It can be waived altogether by franchisors for established franchisees who plan to secure additional franchise restaurants. The decisions pertaining to franchise fees and related terms rest solely with the franchisor. The initial franchise fee is payable in a lump sum and must be paid within a set period of time after signing the franchise agreement.

Some franchisors charge an application fee, which is due when a prospective franchisee's final application is approved, after completing initial screening and interviews. The application fee is a one-time nonrefundable fee designed to cover administrative expenses related to the review and screening of the application. For first-time franchisees, this fee is normally applied to the franchise fee. The application fee is waived for franchisees who are in good standing and would like to develop additional franchise units. A site reservation fee is charged by some franchisors to reserve an area for the approved franchisee. The application fee and site reservation fee become the initial franchise fee when a franchisee is able to find a site and a franchise is available and granted by the franchisor.

Royalty Fee

The royalty fee, as the name indicates, is the royalty payable to the franchisor on a regular basis for securing rights of franchising. This fee is payable when the restaurant is in operation and is based on sales. It varies from franchisor to franchisor and is usually based on the percentage of gross sales. *Gross sales* refers to all revenues related to the franchise business, excluding taxes and amounts received for nonfood items, such as those used for promotional campaigns. It also does not include any discounts deducted from the prices charged. Royalty fees range from 3 to 7 percent based on the gross sales.

Some franchisors also set certain minimum fees, such as the greater of 4 percent of gross revenues and a minimum of $600 per month. This minimum royalty fee may be subject to adjustment and may be based on the Consumer Price Index. It is payable weekly, monthly, or at other set intervals by the franchisor. It can also be based on a scale of percentages that change at a set period of time, such as every five years during a twenty-year period. Gross sales are recorded by the franchisee and the set percentage amount sent to the franchisor when due. The franchisor may also conduct an audit; if there is understatement of gross sales, an audit fee may be charged in addition to making up for the underage. Also, interest may be charged on any payments that are overdue.

Advertising Fee/Fund

This fee is payable to the franchisor and is earmarked for advertising and promotion by the franchisor. The advertising fee is based on gross receipts and is usually placed in a fund used for national and regional advertising and promotion. The franchisor also contributes to the fund for company-owned stores.

Fees for local advertising may be charged separately. Some franchisors give the option of local advertising to franchisees. A grand opening advertising fee is separate from the regular advertising fee and based on the extent of regional promotion needed. The advertising fee is also referred to as *advertising* or *marketing fund*. The national advertising fee ranges from 2 to 4 percent based on gross sales. The local advertising fund ranges from 3 to 4 percent. This fee is subject to increase based on cost increases in advertising and promotion and the media selected. Because the advertising fee has been the cause of misunderstanding or discontent among franchisees, many franchisors are setting maximum limits that can be charged—for example, 6 percent of gross sales with a maximum of $600 and a minimum of $400 per restaurant, payable every month.

In some regions, advertising cooperatives (co-ops) are formed and the advertising fund decided by a vote of its members. An advertising co-op fee is charged to the franchisee. The national and/or local co-op advertising fee may be applicable.

Development Fee

A development fee is charged by some franchisors for further development of the franchise units. This can be a fixed fee per unit or based on a sliding scale. Some franchisors provide incentive programs in areas targeted for development. In such cases, the franchise and royalty fees are considerably reduced. In cases where more than one unit is to be developed, there is a sliding scale whereby the development fee is reduced. Also, some franchisors give franchisees the option to pay a lower development fee but an increased royalty fee, as shown in the table below:

Options	Development Fee	Franchise Fee	Royalty Fee
Standard Fee	$10,000	$20,000	5%
Variable Fee	$5,000	$5,000	7%

A development agreement must be signed by those franchisees who would like to develop one or more restaurants. The development fee is charged per restaurant and varies from $5,000 to $10,000 per unit. Some franchisors charge the same development fee regardless of how many units are involved and, thus, this fee is in addition to the initial franchise fee. The development fee also varies for nontraditional locations such as convenience stores, grocery stores, amusement parks, sports arenas, service stations, hospitals, schools, truck stops, transportation facilities, and retail stores. Incentive options are added as an addendum to the franchise agreement. The options are not changeable once the agreement is signed.

Renewal Fee

A renewal fee is charged for the renewal of the contract. Normally this occurs when there is a substantial change in the type of restaurant desired by the franchisee. This fee can be waived by franchisors for franchisees in good standing. The renewal fee can also be based on a percentage of the current franchise fees—for example, 25 to 50 percent of current franchise fees—and is payable on demand to the franchisor. It can be a fixed fee adjusted by the franchisor based on the Consumer Price Index. The renewal fee is payable on signing the franchise agreement for the renewal term.

Transfer Fee

This fee is charged by the franchisor when a transfer of ownership is desired by the franchisee. This transfer requires prior approval by the franchisor. The transfer fee is paid by the new franchisee. If the transfer is to an existing franchisee, this fee can be waived or reduced by the franchisor. Also, this fee is not charged for transfers to corporations formed by franchisees for the convenience of ownership.

Training Fee/Costs

Costs for initial and on going training are partially or wholly paid by the franchisee. The franchisor may charge a fixed fee for training or may base it on the number of participants. In most instances, travel, lodging, and related expenses are borne by the franchisee. Some franchisors charge a fixed training fee that is payable in its entirety at the signing of the franchise agreement. However, the training fee is waived for those who have previously successfully completed the franchisor's training program.

Opening Costs

Costs incurred for the grand opening of the restaurant franchise are also payable by the franchisee. A representative of the franchisor normally assists in the preopening and grand opening activities and the franchisee is expected to bear all related expenses. As with some of the fees mentioned earlier, some franchisors put a maximum limit on opening costs.

Equipment Costs

Franchisors may require that equipment be purchased through their cooperative or through the franchisor. This is done in order to maintain uniformity of quality. For some franchisees this can be a good deal, as benefits from large-quantity manufacturing/purchase can be passed on to the franchisees. However, this can become a sore point if similar equipment is available at a lower cost than that charged by the franchisor.

Although the fees charged by franchisors are nonrefundable and often nonnegotiable, for promotional purposes and to develop units rapidly, various marketing incentives are provided, which may include rebates, credit against initial franchise fees, and limited royalty waivers. These offers can also be extended to sell franchises at nontraditional locations and to qualified franchisees who would like to develop or convert one concept to another.

INITIAL INVESTMENT

Two of the most common questions asked by franchisees is how much is the initial investment for the franchise and what are the estimated sales and financial projections. This is also the most misunderstood and underestimated aspect of franchising. Tables 8.1 and 8.2 are shown as examples of calculation of estimated sales projections and financial projections. These are given as a worksheet that a potential franchisee can use in performing calculations. It should be clearly understood that franchisors cannot guarantee or promise any financial gains. Table 8.3 describes in detail the initial investment pertaining to the pre-opening costs and for the land or building a restaurant. All estimates are based on averages and are shown only as an example. The costs may vary based on a number of factors, such as the type of restaurant, the style of service, location, and local regulations.

The items listed in Table 8.3 are discussed below. The figures shown are given only to illustrate the categories of expected expenses, which may vary to a wide extent based on individual and specific circumstances.

1. The initial franchise fee is required by the franchisor and is based on the duration of term for which the franchise agreement is signed, which may be from five to twenty years. This fee does not include any development fee or any other fees that may be required to be paid in advance.

TABLE 8.1   ESTIMATED SALES PROJECTION

Item	Estimate	Description
(1) estimated average weekly customer count		Customer count estimates can be derived by estimating the number of customers at a similar franchise restaurant at a similar location.
(2) estimated average daily customer count		Divide (1) by 7.
(3) number of days projected for the restaurant to be open each year		Subtract from 365 the days when the restaurant will not be open, such as during holidays.
(4) annual projected customer count		(2) × (3).
(5) average check expected	$	This depends on the type of menu. Averages can be calculated by selecting the most popular items.
(6) annual projected sales	$	(4) × (5).

2. The required training program and its conditions are outlined by the franchisor. Costs incurred may include salaries for those attending the training program, travel, and lodging expenses. These costs depend upon the distance to the training location, mode of transportation, and accommodations selected. There may be other training fees required to be paid to the franchisor. Franchisees are required to arrange transportation, lodging, food, and incidental expenses for themselves as well as for designated employees. Salaries and benefits for these employees are also payable

TABLE 8.2 FINANCIAL PROJECTIONS

Item	Estimate	Description
(1) projected annual sales	$	From the sales projection worksheet.
(2) cost of goods	$	(1) × cost of goods sold % [this is normally 35–40%].
(3) labor costs	$	(1) × cost of labor %.
(4) royalty fees	$	(1) × royalty fees %.
(5) other fees	$	(1) × other fees %.
(6) operating costs	$	(1) × operating costs %.
(7) depreciation/ amortization	$	Equipment depreciation is calculated by the cost of equipment and installation divided by the expected lifetime of the equipment.
(8) profit contribution	$	(1) − (2) − (3) − (4) − (5) − (6) − (7) is equal to profit.

by the franchisee. The number of employees to be designated for training is determined by the franchisor.

3. Utility and phone costs are variable and depend on location. These costs may be less for established business operations than for new ones. Similarly, the cost of business licenses varies widely based on state, county, and local agencies' requirements. Deposits may or may not be necessary, depending on location. Also, most deposits are refundable after a period of time, if not on the discontinuation of service. Interest is paid on deposits by some utility companies. Some states require deposit for sales tax.

4. Having adequate insurance coverage for any type of business is essential. This is one of the costs for which there should be no cutting of corners. Major insurance policies that should be secured at the time of pre-opening include

TABLE 8.3 ESTIMATED PRE-OPENING COSTS FOR A FRANCHISE RESTAURANT

Items	Estimated Costs	Payment Method	When Due
1. initial franchise fee	$20,000	lump sum	upon signing the franchise agreement
2. travel and lodging expenses during training	$5,000–$10,000	as incurred	as arranged
3. deposits for utilities, business licenses, and sales tax	$500–$10,000	as incurred	as arranged
4. insurance, security deposits, and other prepaid expenses	$3,000–$5,000	as incurred	as arranged
5. initial inventory required for operation (food and supplies)	$10,000	as incurred	as arranged
6. costs for labor (crew and management), food, and training during pre-opening	$10,000	as incurred	as arranged
7. uniforms	$3,000	as incurred	on receipt
8. professional fees	$10,000	as incurred	as arranged
9. advertising (grand opening)	$5,000–$10,000	as incurred	as arranged
10. miscellaneous expenses	$3,000	as incurred	as arranged
TOTAL ESTIMATED PRE-OPENING COSTS	$69,500–$91,000		

liability insurance, worker's compensation insurance, property insurance, and medical insurance. Other insurance and employee benefits may be necessary. Insurance premiums are payable in advance in installments arranged by the insuror. Insurance premiums depend on geographic location, projected sales, projected labor costs, past insurance history, and extent of coverage limit desired. Franchisors may have minimum requirements for insurance purposes, which should be met per the agreement. For finances secured for the purchase of land, property, and equipment, additional insurance may be required by the lenders. If the real estate is leased, the first one or two rental payments may fall within the pre-opening period.

5. Initial inventory of food and supplies is necessary for the pre-opening period. This inventory is obviously larger than that expected after the opening of the restaurant. These costs vary with the type and amount of goods purchased. It should be noted that there are different varieties of goods from which the franchisee can select items for inventory.

6. Before the opening of the restaurant, employee training and trial runs of the operation are necessary. Costs incurred during training, such as for wages, payroll tax, food, and other expenses, should be taken into consideration.

7. Uniforms for employees are required by the franchisors and can be purchased from either the franchisor or franchisors' designated vendor(s). These uniforms should meet the standards set by the franchisors. Uniforms normally carry the trademark, which is the franchisor's property.

8. The professional help of attorneys, accountants, and consultants is needed for incorporating the business, partnership documents, reviewing franchise documents, arrangement for accounting, and other tax matters. Costs incurred for such services are estimated in this item.

9. In-store and local advertising may be necessary during the pre-opening period of the restaurant. Advertising expenses for point-of-sale displays and merchandising should be taken into consideration. Most of the supplies and technical assistance needed for this purpose are provided by the franchisor.

10. Miscellaneous costs may be associated with accounting a change fund for the restaurant, banners and decorative items, and other supplies for use during the pre-opening period only.

As evident from the discussion, the initial investment costs listed do not include the costs associated with land, real estate, renovations, leasehold improvements, or equipment for the restaurant. A listing of the estimated costs for such expenses is shown below. Again, this is only an estimate of expenses that vary widely based on numerous conditions.

As evident from Tables 8.3 and 8.4, the total expenses of opening a franchise restaurant may easily reach half to three quarters of a million dollars for a single unit. These figures should be taken into account when figuring out the return on the investment. Descriptions associated with the items listed in Table 8.4 follow.

1. Real estate for a restaurant can be purchased or leased, based on which costs vary significantly. The cost of land, lease expenses, and building varies with the location of the restaurant. Metropolitan areas are more expensive than nonmetropolitan areas. Also, these costs vary considerably with the design and type of restaurant. Freestanding restaurants

TABLE 8.4   AN EXAMPLE OF ESTIMATED COSTS FOR REAL ESTATE, SITEWORK, BUILDING, AND EQUIPMENT

Items	Estimated Costs
1. land and real estate	$80,000
2. sitework and renovations	$150,000
3. building costs and remodeling	$200,000
4. miscellaneous construction	$30,000
5. equipment, utensils, and signs	$200,000
TOTAL (ESTIMATED COSTS FOR LAND, BUILDING, AND EQUIPMENT	$660,000

are more expensive than smaller kiosks or drive-through restaurants. For purchasing a site for restaurants, the square footage required by the franchisor should be considered. Normally a freestanding restaurant in a suburban location requires from 20,000 to 30,000 square feet of land in addition to adequate parking facilities. Nontraditional restaurants may require less space and may range from 1,000 to 3,000 square feet. If the space is leased, the lease costs vary from $2,000 to $4,000 per month.
2. Work at the site depends on the conditions of the land and access to utilities such as electricity, gas, water, fire hydrants, and drainage. Sitework may be needed for building or repairing pavements, curbs, landscaping, signage, grading, drainage, dumpster sites, and loading facilities. Some of these facilities may already be in place, thereby saving their costs.
3. The construction costs for the restaurant depend on contractor and labor costs in the region. Services provided by most franchisors include architectural and engineering services. Extra costs may be charged by the franchisors for providing these services during the construction of the restaurant. The advantage of utilizing the franchisor's services are that they have a tested plan, may have ready-to-build materials, and meet all the specifications required by the franchisor. Costs for securing building permits and inspections should be considered. The costs for improvements on existing facilities vary on the extent of work needed to meet the franchisor's specifications. Remodeling may be needed to change the exterior as well as the interior of the restaurant in order to meet the image requirement of the particular franchise. Certain items of furniture, fixtures, equipment, signage, and small wares may have to be purchased through the franchisor. These items are listed in a catalogue or manual provided by the franchisor.
4. Miscellaneous expenses incurred for building include architectural and engineering fees, utility hook up costs, environmental clean up costs, and other regulatory fees. These

charges depend on the site and the amount of work needed. More emphasis is being placed on the environmental problems and any site selected for restaurants should be thoroughly studied for adverse environmental impact. Also, all local regulations should be met. Sensitivity of the local population toward development should also be considered, as the success of a business depends on the cooperation of the community.

5. Franchisors require that equipment, appliances, and utensils be purchased from them or their approved vendors. This is done in order to maintain standards of quality in the preparation of menu items, adaptability to layout plans, and the image of the franchise. Buying the equipment through the franchisor can also help in reducing maintenance problems as well as the initial cost of the equipment. This is because the franchisor's representatives are familiar with the equipment; in addition, bulk buying of materials reduces costs — savings that can be passed on to franchisees. Equipment costs depend on the type of restaurant selected. Extra costs may be incurred for selecting a play area for children, exterior decorations such as light and signage, and type of seats and seating arrangement.

The above investment costs are given as examples. Franchisors provide details of the costs to prospective franchisees. Some of the expenses listed may not be applicable and other expenses may arise. In the case of an existing restaurant earmarked for renovation, many of the listed expenses may not be incurred.

COST CONTROL IN RESTAURANT OPERATIONS

Cost control is an essential function of good restaurant management. Several types of costs are incurred in any restaurant operation, the most important of which are food and labor. The cost of food may be determined by the inventories as follows:

Value of opening inventory (for a specific period)
+ food purchases = total available inventory − closing inventory = cost of food consumed (for specific period)

The food cost percentage is the food cost expressed as the percentage of food sales and can be calculated as:

$$\text{Food cost (\%)} = \frac{\text{Food costs}}{\text{Food sales}}$$

Food cost percentage is used as a budgeting tool and for comparative evaluations of financial statements. The food cost percentage varies with the type of foodservice operation. It indicates the amount spent for food out of the total food sales. Similarly, labor costs may be calculated as:

$$\text{Labor cost (\%)} = \frac{\text{Cost of labor}}{\text{Sales}}$$

Food and labor costs are dependent on several factors, including the type of restaurant operation, style of service, menu, and location. The cost of individual menu items may be calculated by using the cost of each item used in the recipe. For items that are used in very small quantities, such as spices and seasonings, a fixed cost of about 1 percent may be calculated.

PROFIT AND LOSS STATEMENT

The most common financial statements used by any type of foodservice operation are the profit and loss statement and the balance sheet. The profit and loss statement, as its name indicates, is a statement of the status of the account showing profit and losses. An example is shown in Figure 8.1. The first information to be recorded includes food and beverage sales and their totals. Similarly, the cost of food and beverages is calculated.

SALES
  Food .... $ .... %
  Beverages .... $ .... %

  Total Food & Beverage Sales .... $ .... 100.00 %

COSTS
  Food .... $ .... %
  Beverages .... $ .... %

  Total Cost of Sales .... $ .... %

GROSS PROFIT
  Food .... $ .... %
  Beverages .... $ .... %

  Total Gross Profit .... $ .... %

OTHER INCOME
  Description .... $ .... %
  TOTAL INCOME .... $ .... %

CONTROLLABLE EXPENSES
  Payroll .... $ .... %
  Employee Benefits .... $ .... %
  Employees' meals .... $ .... %
  Direct operating expenses .... $ .... %
  Music and entertainment .... $ .... %
  Advertising & promotion .... $ .... %
  Utilities .... $ .... %
  Administrative & general .... $ .... %
  Repairs & maintenance .... $ .... %

  TOTAL CONTROLLABLE EXPENSES $ .... %

PROFIT BEFORE RENT $ .... %
RENT OR OCCUPATION COSTS $ .... %

PROFIT BEFORE DEPRECIATION $ .... %

DEPRECIATION $ .... %

PROFIT BEFORE INTEREST $ .... %
INTEREST EXPENSE $ .... %

PROFIT BEFORE INCOME TAX $ .... %
INCOME TAX $ .... %

NET PROFIT $ .... %

FIGURE 8.1 Typical contents of profit and loss statement

Financial Aspects of Franchising

Gross profit is the amount after the cost of food and beverages is deducted from sales. This gives the total gross profit for the restaurant operation. This is the gross profit and does not take into consideration any expenses.

Other income includes income from sources other than food and beverages, such as vending machines, cigarettes, and newspaper sales.

Total income is the gross profit plus other income.

Controllable expenses, as the name implies, are those expenses that can be controlled. They are based on management decisions and are directly related to the efficiency of the restaurant's operation.

Payroll includes all salaries and wages.

Employee benefits include expenses for medical insurance, compensation, and other benefits paid for the employees.

Direct operating expenses are those items directly related to services offered to the customers such as centerpieces, candles, silverware, and napkins.

Music and entertainment expenses are included if these kinds of services are provided.

Advertising and promotion expenses include all kinds of advertising, promotion, and discount expenses.

Utilities expenses are those spent on utilities from all sources.

Administrative and general expenses are overhead expenses not related to services offered to the customers, such as office supplies, postage, and telephones.

Repairs and maintenance expenses include all types of repairs and maintenance expenditures.

Total controllable expenses include all controllable expenses and are sometimes used as an indicator of management efficiency, as all these items may be controlled by the management.

Profit before rent is calculated by subtracting total controllable expenses from the total income. It is also referred to as *operating profit,* as it indicates the profits made by the operation before calculating other costs.

Rent or occupation costs include rent, occupation costs, real estate taxes, and insurance.

Profit before depreciation is calculated by subtracting rent or occupation costs from the profit before rent.

Depreciation is the calculated cost of expected or actual depreciation (wear and tear).

Profit before income tax is calculated by subtracting the depreciation cost from the profit before depreciation.

Net profit is the total net profit derived after subtracting income taxes. This is the total profit after taking into account all costs and expenses.

BALANCE SHEET

An example of a balance sheet is given in Figure 8.2. A balance sheet is the balance of assets versus liabilities and capital. It gives an overall picture of the financial position of any restaurant operation. Some of the terms used in balance sheets are defined below:

Current assets are those assets that can be converted into cash in a relatively short period of time. They commonly include cash on hand, accounts receivable, and inventories.

```
                              ASSETS
CURRENT ASSETS
   Cash on Hand ........................ $
   Cash in Bank ......................... $
                                          ─────────────
   Total Cash ................................................... $

   Accounts Receivable
      Customers .......................... $
      Credit Cards ....................... $
      Employees ......................... $
      Other ................................ $
                                          ─────────────
      Total Receivable ........................................ $

   Inventories
      Food ................................. $
      Beverages .......................... $
      Supplies ............................. $
      Other ................................ $
                                          ─────────────
      Total Inventories ....................................... $
   Prepaid Insurance, Taxes, etc. .................................. $
                                          ─────────────
      Total Current Assets ................................. $ _____
Fixed Assets
   Land .................................... $
   Building ............................... $
   Depreciation (deduct) ............ $_____
   Amount ................................................................ $_____
   Furniture, Fixtures & Equipment ......... $
   Depreciation (deduct) ............ $_____
   Amount ................................................................ $_____
   Leasehold Improvements .................... $
   Depreciation (deduct) ............ $_____
   Amount ................................................................ $_____
   Operating Equipment—China,
      Silver, etc. ............................ $_____
      Total Fixed Assets ................................. $_____

TOTAL ASSETS                                                      $_____

                    LIABILITIES AND CAPITAL
CURRENT LIABILITIES
   Accounts payable—trade ............... $
   Accrued taxes payable ................. $
   Accrued expenses payable ............ $_____
   Total Current Liabilities ................................... $_____
EQUIPMENT AND CONTRACTS PAYABLE ................... $_____
LONG-TERM LOANS ...................................................... $_____
   Total Liabilities .................................................. $_____
CAPITAL (OWNERS' EQUITY) ................. $_____
PROPRIETORS' ACCOUNT ...................... $_____
TOTAL NET WORTH ................................................. $_____
TOTAL LIABILITIES AND CAPITAL ................................ $_____
```

FIGURE 8.2   An example of a balance sheet

Fixed assets are those of a permanent nature and that cannot be converted to cash while the business is in operation. Examples of fixed income include furniture and land. Their depreciated values must be calculated for inclusion in the balance sheet.

Current liabilities are obligations that will become due within one year of the balance sheet date and include such items as short-term loans and taxes collected from customers and payable to the government.

Fixed liabilities are equipment contracts payable or notes payable on a long-term basis. They are not payable within one year of the balance sheet date.

Net worth includes invested capital and earnings retained in the business at the balance sheet date. If a restaurant operation is operated as a partnership, it is preferable to show each partner's net worth separately on the balance sheet.

It should be noted that every item must be converted into dollars in order to be placed in the financial statements.

RATIO ANALYSIS

In order to interpret and analyze financial statements, various analyses and ratios are calculated; the most important ones are described below:

Liquidity ratios help in analysis of an operation's ability to meet short-term obligations as and when they become due. The *current ratio* is probably the most commonly used ratio in the restaurant industry. This ratio expresses the relationship between the total current assets and the total current liabilities. The *acid test ratio* or *quick ratio* is another way of comparing current assets and liabilities. *Accounts receivable as a percentage of total revenue* represents the portion of revenues that have not been converted into cash and therefore are not available for payment of current obligations. *Accounts receivable turnover* shows the rapidity of the conversion of accounts receivable to cash. The calculations used for liquidity ratios are:

$$\text{Current ratio} = \frac{\text{Current assets}}{\text{Current liabilities}}$$

Acid test ratio (quick ratio)

$$= \frac{\text{Cash + accounts receivable + marketable securities}}{\text{Current liabilities}}$$

or

$$\frac{\text{Current assets} - \text{inventory}}{\text{Current liabilities}}$$

Accounts receivable to total revenue ratio

$$= \frac{\text{Average accounts receivable}}{\text{Total revenues}}$$

Accounts receivable turnover

$$= \frac{\text{Total revenue (total credit sales)}}{\text{Average accounts receivable}}$$

Solvency refers to the ability of an operation to meet its debt obligations when they become due, including principal and interest on long-term borrowing. Solvency and leverage ratios are therefore designed to calculate the extent to which an operation is able to meet its debt and how much leverage exists for this possibility. Equations used for calculating solvency and leverage ratios are:

$$\text{Solvency ratio} = \frac{\text{Total assets}}{\text{Total liabilities}}$$

$$\text{Debt to total assets ratio} = \frac{\text{Total liabilities}}{\text{Total assets}}$$

$$\text{Debt-equity ratio} = \frac{\text{Total debt}}{\text{Total equity}}$$

Number of times interest earned ratio

$$= \frac{\text{Net profit before income taxes and interest expense}}{\text{Interest expense}}$$

Activity ratios show how effectively the assets of a restaurant operation are being utilized. They are also indicative of the efficiency of management and may be calculated using one of the following methods. It should be noted that each method measures a different parameter, which should be considered when comparing data.

$$\text{Food inventory turnover} = \frac{\text{Total cost of food sold}}{\text{Average food inventory}}$$

$$\text{Beverage inventory turnover} = \frac{\text{Total cost of beverage sold}}{\text{Average beverage inventory}}$$

$$\text{Fixed asset turnover} = \frac{\text{Total revenues}}{\text{Total fixed assets}}$$

$$\text{Accounts receivable turnover} = \frac{\text{Total sales}}{\text{Average accounts receivable}}$$

Profitability and *rate of return ratios*, though not all applicable to all types of restaurant operations, are measures of profitability. They may be calculated as follows:

$$\text{Return on owner's equity} = \frac{\text{Net profit after income taxes}}{\text{Average stockholder's equity}}$$

Return on assets

$$= \frac{\text{Net profit before interest and income taxes}}{\text{Total average assets}}$$

$$\text{Net return on assets} = \frac{\text{Net profit after taxes}}{\text{Total average assets}}$$

$$\text{Profit margin} = \frac{\text{Net profit after income taxes}}{\text{Total revenue}}$$

$$\text{Operating efficiency ratio} = \frac{\text{Gross operating profit}}{\text{Total revenue}}$$

Operating ratios provide important information pertaining to daily restaurant operations as well as to the efficiency of management. These ratios measure are helpful in assessing various operating parameters of an establishment. There are several ways in which these ratios may be calculated, some of which are given below:

$$\text{Average restaurant check} = \frac{\text{Total food sales}}{\text{Number of food covers sold}}$$

$$\text{Food cost (\%)} = \frac{\text{Cost of food sold}}{\text{Food sales}}$$

$$\text{Beverage cost (\%)} = \frac{\text{Cost of beverage sold}}{\text{Total beverage sales}}$$

$$\text{Sales per seat} = \frac{\text{Net sales}}{\text{Number of seats}}$$

$$\text{Operating ratio} = \frac{\text{Net profit}}{\text{Net sales}}$$

$$\text{Net profit to net ratio} = \frac{\text{Net profit}}{\text{Net worth}}$$

$$\text{Management proficiency ratio} = \frac{\text{Net profit after taxes}}{\text{Total assets}}$$

Although ratio analysis is a powerful analytic tool in evaluating financial ability, management effectiveness, operating results, activity reports, and as an aid to management in decision making, it should be used with caution because it has the following limitations:

- Ratios are mathematical calculations and do not represent the human side of management. They are good only as far as the comparisons are meaningful.
- Because ratios are based on mathematical calculations, any error or difference in the reporting system offsets the basis of comparison.
- The timing of financial transactions is important in comparing these ratios. Conditions change rapidly in any type of financial business and this should be taken into consideration.

If properly interpreted, however, ratios can be valuable tools in the hands of management.

Restaurant Study 13

International Dairy Queen, Inc.

In 1938, the United States was emerging from the Great Depression. People were working again and families had money to spend on recreational activities and extras like eating out.

At this time, J. F. McCullough, owner of the Homemade Ice Cream Company in Green River, Ill., was experimenting with the idea of creating a new frozen dairy product. McCullough felt ice cream tasted better when it was soft and dispensed fresh from the freezer, not frozen solid.

To test his theory with the public, McCullough and his son and business partner, Alex, approached one of their best customers, Sherb Noble, with the idea of holding an introductory sale of the soft ice cream product at Noble's ice cream shop in Kankakee, Ill. Noble agreed to host the event, and the first "All the Ice Cream You Can Eat for Only 10 Cents" sale was held August 4, 1938.

The response was overwhelming! Lines of customers stretched for blocks and more than 1,600 people were served the soft ice cream in two hours. Noble feared that the push of the crowd would break in the window in the front of his store. The success of this sale, and a similar sale held two weeks later, convinced the McCulloughs that their new product was a big hit.

The next challenge facing the McCulloughs was that of finding a reliable freezer to dispense the soft serve product. Regular ice cream freezers of that day could dispense the product at the correct temperature but were not capable of maintaining the product at the perfect 23 degrees Fahrenheit. The problem was solved when the McCulloughs answered a newspaper advertisement placed by Harry Oltz, who had received a patent for a continuous-type freezer in 1937. The McCulloughs and Oltz soon began working together, and in 1939, they signed an agreement that gave the McCulloughs the rights to produce the freezers and exclusive rights to use them in the western United States. Oltz received the exclusive rights to use the freezers in the eastern part of the country, along with royalties based on gallons of the soft serve mix processed by all of the freezers. This agreement launched the business that was to become known as the *Dairy Queen* system.

By the summer of 1940, the McCulloughs were ready to open the first store and sell the soft serve product. The owner of this first store was none other than Sherb Noble, who had hosted the first "All You Can Eat" sale. The store was located in Joliet, Ill., and opened for business June 22, 1940.

When choosing a name for the new store, J. F. McCullough looked for a way of describing the freshness of the new product the store would sell. Having been in the dairy business all his life, McCullough considered the cow to be the queen of the dairy industry. And, since he believed his soft serve creation was the closest thing to real dairy freshness, he decided on the *Dairy Queen* name.

But, even as the celebration of a successful new store continued, a war was being fought in Europe. When the United States entered World War II in December 1941, there were fewer than 10 *Dairy Queen* stores. During the war years, that number remained constant. The few stores that were open found it difficult to locate the dairy products needed to keep the business operating. Doors were open just as long as the rationed products held out; then they closed for the rest of the month.

The menu in these first *Dairy Queen* stores was a far cry from the hot foods, cold drinks, and tempting novelties found today. The menu in a 1940s *Dairy Queen* store consisted simply of the soft frozen dairy treat served in sundaes, take-home pints and quarts, and two sizes of the trademarked *Cone with the Curl on Top*.

As the war ended, the United States was ready to explode with four years of built-up consumer demand. The growth in popularity of the soft serve product and the number of *Dairy Queen* stores exploded, as well. A system that began with that single store in Joliet, Ill., in the summer of 1940, and had only 100 stores in 1947, grew to a nationwide system of 1,156 stores by 1950.

Contributing to the rapid growth of the system was a network of independent businesspeople called Territory Operators. These Territory Operators spread out across the country and were granted the rights by the McCulloughs and Oltz to sell *Dairy Queen* franchises in a specific geographic area, usually a county or state. Territory Operators still play an important role in the *Dairy Queen* system today. Little did they know at the time, but through these agreements the McCulloughs and Oltz were laying the groundwork for the system of franchising that many fast-food organizations operate under today.

In an effort to maintain uniformity and stability in the rapidly growing *Dairy Queen* system, operators joined together in 1948 to form the Dairy Queen National Trade Association. The task of the trade association was to develop systemwide advertising efforts, provide store operators with group purchasing power, and standardize operations nationwide.

With the 1950s came air-conditioning, television, and growing families. These factors kept people at home in the evenings, and one way the *Dairy Queen* system responded was to expand the menu. Such all-time favorites as the Banana Split, the *Dilly* Bar, the *DQ* Sandwich, and the *Buster* Bar each made their debuts in *Dairy Queen* stores during this decade.

Also during the 1950s, a Territory Operator in Georgia began experi-

(a)

(b)

An exterior view of a Dairy Queen restaurant (courtesy American Dairy Queen Corp.)

menting with the addition of hot food to the menu. This development eventually led to the *Brazier* food line now offered in many *Dairy Queen* stores.

These menu additions in the 1950s contributed to sustained, dramatic growth for the *Dairy Queen* system. By the time the 1960s arrived, there were 3,000 *Dairy Queen* stores in the United States, Canada and 10 other countries.

To launch this new decade, a unique logo was developed to represent the *Dairy Queen* name. The red ellipse *DQ* logo is now one of the most recognizable corporate symbols in the world.

The 1960s also brought many changes to the operations of the *Dairy Queen* system. The most significant was the pooling of resources by several Territory Operators in 1962, which led to the establishment of International Dairy Queen, Inc.

Among the new products introduced in the 1960s was the noncarbonated, icy beverage known as the *Mr. Misty* drink, which was introduced in 1963. In 1968, the *Brazier* food line became available systemwide.

By the 1970s, the *Dairy Queen* system had become firmly positioned as a worldwide favorite. In 1970, IDQ was acquired by a group of investors who ensured financial stability as the system moved onward. The following year saw the first trading of IDQ stock in the over-the-counter market.

That same year, 1971, the *Dairy Queen* system added a new spokesman, familiar to millions around the world, to present the message that *Dairy Queen* stores are a fun place for families to go. "Dennis the Menace" has been featured in television commercials, on premium items, and on *Dairy Queen* packaging since that time.

Aggressive international expansion also continued through the 1970s. *Dairy Queen* stores opened in Japan, Hong Kong, Iceland, Panama, Puerto Rico, and the United Arab Emirates. The early 1980s brought the *Dairy Queen* system its first year with sales over $1 billion. This, however, was just the start of significant milestones for the *Dairy Queen* system in the 1980s.

In 1983, *Dairy Queen* stores made the commitment to become one of the largest sponsors of the Children's Miracle Network Telethon. Since joining the telethon effort, *Dairy Queen* stores have raised millions of dollars for a network of more than 160 hospitals for children across North America.

New products joining the *Dairy Queen* menu in the 1980s included the Full Meal Deal in 1980, frozen cakes and logs in 1981, *Queen's Choice* Hard Ice Cream in 1983, and the *DQ Homestyle* Burger in 1984.

In 1985 the most successful product introduction of all time took the *Dairy Queen* system by storm. The product was the *Blizzard* Flavor Treat. More than 200 million were sold in 1985 alone.

Significant milestones of the 1980s were capped by the grand opening celebration of the 5,000th *Dairy Queen* store in November 1987.

With the *Orange Julius* and *Karmelkorn* systems joining the IDQ franchise family in the 1980s, the *Treat Center* concept was created for high traffic areas such as shopping malls. The *Treat Center* design joins together *Orange Julius* drinks, *Karmelkorn* popcorn snacks, and *Dairy Queen* soft serve under one storefront. The first *Treat Center* unit opened in 1987.

CHAPTER 9

Franchisor–Franchisee Relationships

In 1990 the Congressional Committee on Small Business published a review of the franchisor-franchisee relationship. The conclusion of the report states:

> While problems remain to be resolved involving disclosure of franchising earnings claims, the critical issues of franchising no longer center on fraud and misrepresentation in franchise sales. The focus of state regulatory activities, state legislation and legal actions has shifted during the past decade to problems involving abuses within ongoing franchise relationships. Congress has not addressed these "relationship" issues in any comprehensive manner since the late 1960s. Such consideration is long overdue.

The above conclusion is representative of problems between franchisors and franchisees. As noted in earlier chapters, franchising is a unique symbiotic relationship that must function smoothly in order to be successful. Franchising relationship are often compared to marriage. There is a falling in love period, followed by a marriage and a honeymoon period. And, yes, there are rocky relationships and divorces. The major friction points between franchisor and franchisee are described in the following sections. The cause of friction seems to be money; franchisors claim they are not getting royalties or franchisees claim they are not getting services for their money. Because franchising is a mutual relationship, one of the clear ways of maintaining that mutuality is for both parties to be honest and fair with each other. The system fails if understanding lacks in any way. The success of the franchisee-franchisor relationship is achieved in the same way as any other successful relationship—by mutual understanding. When the relationship is out of balance, when one party feels or perceives that the other is contributing less, the relationship begins to crack and crumble. Franchising should ideally be a win-win situation, with a handsome margin of profitability built into the system.

Friction Points

Friction in the relationship can be traced to both franchisors and franchisees. Sometimes a franchisor get carried away and tries to sell more franchises than the system can handle, to the extent of admitting unqualified franchisees. An overambitious franchisor may not have time to devote to each franchisee's problems or to assure the maintenance of quality. If one franchisee is negligent, the entire system can be affected. Once a franchise agreement is signed, a long-term relationship is begun, and if problem franchisees are allowed into the system, a long-term problem results. The primary function of the franchisor is to maintain the quality of franchise products and services. Cooperation from both franchisees and franchisors is essential for quality maintenance.

Other problems are related to inadequate communication between franchisor and franchisees. Many friction points can be traced to the franchise agreement and its interpretation. Many franchisees feel strongly that the termination and other regulating clauses in the agreement favor franchisors to the disadvantage of franchisees.

Allocation and use of expenses may represent a sore point in the relationship. Many franchisees feel that they are paying more for the benefits than they are getting. Franchisees may not approve of the advertising and promotion plans of the franchisor. At times the advertising does not have any impact in the area where a franchise unit is located; also, some units do not need any advertising, but franchisees still pay a percentage of their gross sales. All of these have an impact on the relationship. To illustrate some of the friction points, their possible causes, and possible solutions, the following examples are given.

FRANCHISEE CONCERNS

Problem: Franchisors are making all the money while franchisees are doing all the work.

Cause(s): Franchisees do not understand the franchise relationship. Franchisors are not concentrating on the franchise. There may be a lack of communication between the franchisor and the franchisee.

Solution(s): Franchisors should explain the relationship prior to the signing of the franchise agreement. There should also be effective communication between franchisors and franchisees in the form of frequent meetings, newsletters, and other publications. Relationships function best when the profitability is shared between the franchisors and the franchisees. Leading franchisors have traditionally cared about the franchisees first and then for themselves. If franchisees can be shown that the percentage of profits shared with the franchisor is fair and reasonable, satisfaction is more likely.

***Problem:* Franchisors are not providing ongoing services that were promised prior to signing the franchise agreement and that were provided during the opening of the franchise.**

Causes: Franchisors are not concentrating on the operational aspects of the franchise. Franchisees' expectations of franchisor's services need clarification.

Solution(s): Franchisors should concentrate on the operational aspects of the franchise. Services promised should be carefully planned and executed. Certain services are pre-opening services and others are ongoing. Some franchisees misinterpret the extent of the pre-opening services as indicative of those to be provided on an ongoing basis. Services to be provided must be explained clearly in writing as well as orally. Frequent reminders during meetings and conferences will help to build confidence and eliminate misunderstanding.

***Problem:* Purchasing requirements are unreasonable, and franchisors are making more profits than necessary by imposing these requirements.**

Cause(s): Franchisors' charges and requirements are unrealistic and need reevaluation. Franchisees do not understand the costs franchisors pay and the markup system.

Solution(s): Franchisors should reevaluate charges, expectations, and need for purchase requirements. If certain markups are necessary they should be explained to the franchisees. The reasons for all franchisor actions should be clear in the minds of the franchisees. To be fair, if possible, choices should be provided to purchase either from the franchisor's agent or any other supplier as long as specifications are met. A comparative assessment of costs and savings should be shown to the franchisees. The proprietary rights and the reason for such purchase requirements should be communicated to the franchisees. If the quality of the product is tied to the purchase of the ingredients or the use of specially designed equipment, this should be made clear to the franchisees.

Problem: **Advertising fees are unreasonable. Advertising and promotions are of little or no advantage to the franchisee.**

Cause(s): Advertising fees and the target advertising markets may not be well structured and suitable. The quality of advertising and promotional materials is poor.

Solution(s): Franchisors should reevaluate fee structures for advertising and plan advertising and promotions that will benefit the franchisees. Adequate planning is required for advertising and promotion, for which external agents and consultants can be helpful. Franchisee advertising and promotion committees are helpful in providing valuable input and assistance in planning.

Problem: **Franchisor's standards and inspection procedures are unreasonable.**

Cause(s): Franchisor's standards are outdated and may not have been tested prior to implementation. Inspection procedures may not be appropriate.

Solution(s): All standards should be reevaluated to assess their applicability. Any standard that is not helping the objectives of the franchisor should be discarded. Inspection procedures should similarly be reevaluated. Franchisees' input in setting standards and inspection procedures should prove valuable.

Problem: **Franchisors accept franchisees who are not qualified and not investigated thoroughly prior to signing the agreement.**

Cause(s): Franchisor's selection criteria are inadequate. Franchisor is too interested in rapid expansion.

Solution(s): Some franchisors are carried away by a wave of expansion. This may be emerge from the need to grow or the need to secure funds. Overenthusiasm can be solved by reducing the pace, but the urgent need for funds is a serious sign of problems. Expansion for the sole purpose of collecting funds runs contrary to the principles of franchising and leads to discontent among franchisees. Careful selection of qualified franchisees is necessary for the success of any franchise system. Financial capability of the franchisees should not be the only criterion. Franchising is a long-term relationship, and the acceptance of unqualified franchisees has repercussions that will last for a long time. Poor performance of a problem franchisee affects the performance of other well-qualified and hardworking franchisees and, finally, the entire franchise system.

Problem: **Franchisor tries to expand too quickly and the system cannot handle the strain.**

Cause(s): Franchisor's expansion plans are unrealistic.

Solution(s): Franchisor should reevaluate the expansion plan. The advisory staff should also be expanded as more franchises are sold. Franchisee selection should not be affected by the expansion, and the maintenance of standards should not be sacrificed. Services provided should not be adversely affected by the expansion plans of the franchisor.

Problem: **Operating procedures are not clear to the franchisees and many questions are not clearly answered even after months of the restaurant being in operation. The franchisor's representative is not available when needed.**

Possible cause(s): The operating manual is not up to date and does not have adequate information. The training program of the franchisor is poor. There is a lack of qualified personnel to provide assistance to the franchisees.

Solution(s): The operating manual should be reevaluated and revised if necessary. Input from the franchisees or a franchisee advisory council should be solicited when revising and updating the operational manual. The training program should also be reevaluated and upgraded. Franchisee input in planning training programs may be desirable. Experienced personnel, possibly former franchisees, should be put in charge of providing assistance to the franchisee. A twenty-four-hour toll-free telephone number should be set up and the franchisor's representative should respond promptly to all queries, whatever the emergency situation. The number of assistance staff should be planned proportionally to the needs of the franchisees. Increasing the number of field representatives and setting up hotline(s) should be helpful.

Problem: **Franchisor's responsibilities are not clear. The franchisor seems to have more control and sense of ownership of the franchisee's business than desirable.**

Cause(s): Lack of information about the franchisor and the principles of franchising available to franchisees.

Solution(s): The concept of franchising and the controls exercised by franchisors should be explained to the franchisees. Communications should be improved by newsletters, brochures, and other materials. The sharing of responsibilities inherent in franchising should be made clear to the franchisees.

Problem: The expectations of the franchisees were not met by the franchisor, either intentionally or unintentionally.

Cause(s): The franchising agreement is not clearly understood.

Solution(s): The franchising agreement is a legal document and should be thoroughly understood and adhered to by both parties. Franchisors should explain all items clearly and should ask the franchisee or his or her legal representative if all clauses are understood. Expectations and responsibilities should be clearly explained. An agreement written in simple language is always an asset. Franchisees should contact other franchisees, even those who are not listed as franchisor references, to verify information and enquire about expectations that were met and those that were unfulfilled.

Problem: Territorial rights are unfair and lead to unfair competition.

Cause(s): The franchise agreement is not clearly understood.

Solution(s): Franchising works by mutual understanding. Territorial rights help franchisors as well as franchisees in expansion. As long as markets are fully explored, both parties can benefit. A franchisor should be considerate of the franchisee's concern. Franchisees should also realize their capabilities and limitations and the reasons for such rights and actions. Contract clarification and mutual understanding will be helpful in finding solutions to such a problem, which will get more complicated if not clarified promptly. Franchisors should realize that overbuilding is not in their best interest.

***Problem:* Falloff in profits blamed by the franchisees on franchisors.**

Cause(s): There can be several causes for this problem. The root may lie in the actions of either the franchisee or the franchisor. This problem may be due to the lack of strategic planning and marketing research by the franchisor or to poor management by the franchisee.

Solution(s): Financial statements and overall status of both franchisor and the franchisee should be evaluated. Financial reports should be analyzed and the problem pinpointed before any solutions are planned.

***Problem:* Franchisor gets distracted by other things and franchisees feel abandoned.**

Cause(s): Franchisor is spread too thin and not concentrating on the franchise system.

Solution(s): Some franchisors make it a point to stay exclusively in the same business as the franchise in order to keep pace with market research and development. If the franchisor is distracted by other investments, it should not be at the cost of the franchisees' needs. In every step a franchisor takes, franchisees should be taken into consideration. Even the long-term research and development designed to benefit the franchisee should not distract the franchisor from immediate concentration on franchisee affairs.

Problem: **The franchisor's training program is inadequate and the follow-up training is insufficient.**

Cause(s): Insufficient training and possibly outdated training methods.

Solution(s): Training methods should be designed to give adequate training to the franchisees. There should be an ongoing program of training that includes additional and updated information. Feedback from the franchisees or their advisory committees is important and helps update the program. Proven teaching and learning principles should be taken into account while planning the training program. Advice from external consultants should be sought if needed.

Problem: **Franchisor's termination and renewal clauses are unreasonable and biased toward the franchisor.**

Cause(s): Franchising agreement needs reevaluation and assessment.

Solution(s): The franchising relationship is based on mutual understanding. A franchisee who has invested a significant portion of his or her life and capital should not be terminated without due cause. Reasons should be clarified and justified, and franchisees should understand the reasons for such termination. It should be understood that it is not in the interest of the franchisor to terminate successful franchisees, whereas problem franchisees deserve termination. All these aspects should be considered because termination can a complicated problem for both franchisee and franchisor. Franchisees should also be aware that because agreements are long term, business conditions may have changed drastically by the time of renewal. Many changes usually have to be incorporated in a contract renewal.

***Problem:* Franchisees do not perceive the franchise concept as distinctive enough.**

Cause(s): Franchisees are too familiar with the concept and possibly losing interest in the franchise.

Solution(s): A franchise relationship begins when the franchisee is interested and excited about the concept and believes that it is promising and unique. After a certain point, franchisees may feel that they know the concept to the extent that it no longer seems unique. To forestall this, the franchisor should keep adding products and services to maintain interest and excitement. Franchisees should also be given opportunities to expand the franchise or to join the corporate staff, if at all possible.

***Problem:* Franchisor fails to be innovative and does not change to suit consumer trends.**

Cause(s): Lack of research and development activities by the franchisor. Lack of communication between franchisor and franchisee.

Solution(s): Franchising is a dynamic business and requires constant research and development efforts. Franchisors should have adequate ongoing research and development plans. Franchisors should recognize that that research and development requires a considerable amount of time. While efforts are going on, franchisees should be made aware of the efforts of the franchisor to remain competitive in the market. Franchisors should remember that most franchisees are entrepreneurs and would rather see things dynamic than static. Innovative efforts are always appreciated by such franchisees.

Problem: **Franchisor does not want to change and is unwilling to accept new systems.**

Cause(s): The franchisor is too rigid to accept any changes.

Solution(s): Franchisors are at times very rigid and do not want to change any system. They may be extremely cautious and not want to change a system that is working. But in order to remain competitive in the long run, some change and risk-taking may be necessary. Franchisors should have careful and calculated long-term strategic plans. Changes do not have to be made for the sake of change only. Feasibility studies and the expensive involved should be considered before making any changes.

Problem: **Franchisees have no influence in matters of corporate strategy, pricing, and policy formulation.**

Cause(s): Lack of franchisee input.

Solution(s): Franchisees are a part of the franchise system and should be consulted in the formulation of strategies and policies. Franchisee advisory committees can help in obtaining franchisee input. Franchisees should be made to understand corporate policies, particularly as to how they benefit franchisees. Franchisees should be involved in pricing decisions, which should be based on sound rationale and reasoning.

Problem: **The franchisor finds business to be easy while franchisees do not.**

Cause(s): Lack of training and information on operations.

Solution(s): Franchisors have developed the concept and have lived with it much longer than have the franchisees. Therefore, franchisors should find it easier. Franchisees need more training and information, especially pertaining to operations.

Problem: **Franchisors have different standards for company-owned units than for franchised units.**

Cause(s): Standards need reevaluation. Inspection procedures are inadequate.

Solution(s): Standards are set for the specific purpose of enhancing business, whether corporate units or franchised units. A consumer probably does not notice or care to notice if a unit is franchised or not. The emphasis should be on uniform standards and procedures for inspection. Franchisors should reevaluate these procedures to implement uniformity. Moreover, company-owned units serve as models or examples for the franchisees and therefore should follow standards even more strictly.

Franchisor Concerns

Problem: **Franchisees are not paying royalties or are refusing to do so.**

Cause(s): Lack of cooperation and understanding; other problems that may or may not be associated with franchise.

Solution(s): Franchisees should be made to understand the franchise relationship and the need for the payment of royalties. The consequences of not paying royalty fees should be explained to franchisees. Other underlying problems should be explored to find reasons for a refusal. It should be demonstrated that both the short- and the long-term success of the franchising system depends on fees paid by franchisees.

Problem: **Franchisees are not following purchase specifications and requirements.**

Cause(s): Lack of understanding of the reasons for purchase requirements.

Solution(s): The franchisor should make clear why it is necessary to adhere to purchase specifications and requirements. Such requirements are essential for the maintenance of quality standards. Information should be provided through multiple communication methods, including franchisee advisory committees.

Problem: **Franchisees are not following instructions and are negligent in maintaining standards.**

Cause(s): Lack of information, improper training, and lack of ongoing communication.

Solution(s): Franchisees should be made aware of the need for maintaining standards and the consequences of not following the standards. Lack of proper training may lead to nonconformance with the franchisor's instructions. Franchisees should be trained and retrained to be made aware of the complexities of the franchise business.

Problem: **Franchisees lack adequate capital at the beginning stages of the business.**

Cause(s): Lack of proper selection of the franchisees.

Solution(s): The major factor to be considered in screening prospective franchisees is their financial capability. Adequate checking and screening should be conducted prior to selection to ensure that enough capital is available, sufficient to sustain the franchise unit until the business is off and running.

Problem: **The Franchisee did not realize that the work involved was so hard and time-consuming. The franchisee is not prepared for the complexities of the business.**

Cause(s): Lack of training.

Solution(s): Franchisees should be trained in franchise restaurants long enough for them to realize the necessary amount of work and time commitment. They should work in all aspects of the operation to gain first-hand experience. Proper training helps prepare franchisees to go ahead with the business or to decide not to enter into an agreement if they feel that this is not the type of business for them.

Problem: **Franchisees are anxious to make changes to help the business.**

Cause(s): Lack of flexibility in franchisor's plans. Lack of communication.

Solution(s): Franchisees are entrepreneurs and so would like to see changes. Franchisors should communicate to them the pros and cons of such changes and listen to their reasoning. A dialogue resolving their anxiety is better than ignoring their pleas.

Problem: **Franchisees resist installation of new systems and other changes. Expense is given as the main reason for the resistance.**

Cause(s): Lack of communication and information related to changes.

Solution(s): Logically, any change that increases costs will be resisted by franchisees. The return on investment and the benefits derived from such changes should be adequately communicated before implementation. Franchisee advisory committees should be involved in such decisions to secure optimum franchisee support.

RELATIONSHIP CONSIDERATIONS

These franchisor-franchisee problems, their possible causes, and solutions are shown as examples. Franchisors should try to anticipate and prevent problems before they become chronic. Satisfaction and dissatisfaction among franchisees should be surveyed regularly. A suggested franchisee survey form is shown in Figure 9.1. The results should be analyzed and problem areas corrected. Major categories included in this survey include financial, legal, training, advertising, quality control, and general information. Such a survey demonstrates the franchisor's concern for franchisees and helps to build a positive image of the franchisor.

Franchisors should make decisions that are in the best interest of the whole franchise system on a long-term basis. Carefully considered decisions help the franchisor as much as the franchisees. Some decisions may not be compatible with the short-term interest of franchisees. Cost may be a major consideration. For example, a franchisor may want to remodel all restaurants for consistency, which an individual franchisee may find unnecessary and incompatible with his or her immediate interest. Franchisors should develop systems that are beneficial to both franchisees and themselves.

Franchisors should realize that franchisees can make significant contributions to their system. Some of the most profitable ideas adopted by leading franchises were initiated by franchisees. In addition to innovative ideas, franchisees also help to maintain a check and balance system. Franchisees are more accepting of a proven system. For example, most franchisees were

For the following please indicate the extent to which the following statements are correct.

5 = Extensively	4 = Greatly	3 = Somewhat		2 = Very little		1 = Not at all

FINANCIAL MATTERS

Initial franchise fee charged is regarded as reasonable by franchisees	5	4		3	2	1
Franchisees find it desirable for the franchisor to maintain a trust fund from which they can borrow	5	4		3	2	1
Fees charged for the services provided are often *too high* for the franchisee	5	4		3	2	1
Franchisees prefer that royalty fees be based on unit performance rather than on the gross sales alone	5	4		3	2	1

LEGAL MATTERS

Franchisees perceive that you often perform according to the legal contract	5	4		3	2	1
Franchisees have had all legal implications made clear to them prior to the signing of the contract	5	4		3	2	1
Franchise contracts are *often* inclined toward the franchisee's interest	5	4		3	2	1
Franchisees receive legal assistance from the franchisor when needed	5	4		3	2	1
Franchise contracts *often* restrict franchisee's choice of purveyors	5	4		3	2	1

TRAINING

Franchises are satisfied with the training program offered by franchisor	5	4		3	2	1
Franchisees are satisfied with training manual	5	4		3	2	1
Franchisees have input into the components of the training manual	5	4		3	2	1
Franchisees do not receive training assistance after the "opening" period	5	4		3	2	1

ADVERTISING

Franchisees are always consulted for any national advertising campaigns	5	4		3	2	1
Franchisors *never* ask franchisees for feedback on national advertising campaign	5	4		3	2	1
Regional advertising should be given more importance than national advertising	5	4		3	2	1
National advertising has *no effect* on the success of the franchisee	5	4		3	2	1

FIGURE 9.1 An example of a survey instrument for evaluating franchisor-franchisee relationships (for franchisor's use)

Promotional coupons are an effective tool of the advertising campaign	5	4	3	2	1

QUALITY CONTROL/STANDARDS

Franchisees are satisfied with the product quality standards set by franchisor	5	4	3	2	1
Sanitation standards are difficult for franchisees to maintain	5	4	3	2	1
The possible consequences of *not meeting* inspection standards are severe for the franchisee	5	4	3	2	1
The quality controller's inspection procedures are acceptable to franchisees	5	4	3	2	1

COMMUNICATION

The most common direction of communication is from franchisor to franchisee	5	4	3	2	1
A franchisee advisory committee helps in reducing the conflicts between franchisees and franchisors	5	4	3	2	1
Communication between you and your franchisee is informal and personal	5	4	3	2	1
If franchisees are satisfied with the present communication system	5	4	3	2	1

GENERAL

Your overall performance is satisfactory to your franchisees	5	4	3	2	1

If the national advertising budget was *optional*, what percentage of the gross sales would your franchisees subscribe?

0%   1%   2%   3%   4%   5%   6%   7%   8%   9%   10%

How often does your quality control inspector visit your franchisees' operations?

___ once a month   ___ once in 2 months   ___ once yearly
___ once a quarter   ___ once in 6 months

Approximately how often do franchisors contact your franchisees?

___ once a week   ___ once a fortnight   ___ once a month
___ once a quarter   ___ once every six months

How would you rank your *overall* relationship with your franchisees/franchisors?

Excellent 5___   4___   3___   2___   1___ Poor

How many units of this franchise do you own?(___)

How long have you been a franchisee of this particular franchise? ___years

Do you have a franchisee advisory committee? Yes___   No___

List one aspect that you feel will help to improve your relationship with your franchisees.

Add any comments on your relationship with your franchisees.

FIGURE 9.1 *(Continued)*

apprehensive when breakfast was added to the menu of certain chains. When it proved a profitable venture, franchisees stood in line to join the system. One way of demonstrating the benefit of innovative changes is to try them first in company-owned units.

Careful screening of franchisees prior to selection is all important. Quality should not be sacrificed at any cost. A weak franchisee can damage the entire system irreparably. The impact of proper training should not be underestimated. A sound training program and effective ongoing communications help to eliminate many of the friction points between franchisors and franchisees.

Franchisors should carefully select the services they provide. Unnecessary services may damage the system as much as inadequate services. As franchisors expand the system, changing needs should be taken into account.

FRANCHISING RELATIONSHIP REGULATIONS

Currently there is no federal law governing franchising relationships or practices, such as termination regulations, renewal regulations, or transfer regulations. State franchise laws are inconsistent with each other, and they may or may not apply to franchising relationships directly. Legislative and regulatory efforts tend to reflect the concerns of franchisees or potential franchisees in two circumstances: (1) problems that arise prior to entering into a franchise relationship, such as deception or impropriety in the presentation, solicitation, or sale of a franchise opportunity, and (2) problems arising within ongoing franchising relationships, such as contract performance and termination or renewal of franchise agreements. While disclosure requirements and procedures both at the federal and state levels address the former situation to a much greater extent, franchising regulations pertaining to the latter set of problems are covered by different regulations.

The dominant problem arising between franchisors and franchisees involves the conditions giving rise to the severance of a franchise relationship by the franchisor's power to terminate, decline to renew, or deny the franchisee the right to sell or transfer a franchise. Franchisees have sought legal protection for many of the problems arising from this situation. Franchisors contend that such power is essential to the efficient operation of a franchise system. Legal actions designed to restrict this authority are considered by franchisors as tantamount to sanctioning noncompliance, incompetence, and uncompetitiveness.

According to the report of the Committee on Small Business of the House of Representatives, submitted to the 101st Congress of the United States,

> The critical issue for regulation in this area is one of finding a balance between the right of franchisors to manage their franchise systems using their best business judgment, and the right of franchisees to receive the full rewards of their effort and investment. Regulation must seek to facilitate cancellation of a franchise wherever such action is warranted. The franchising relationship is one based on mutual trust and a shared desire to work together toward common goals. If this trust is missing, or if a "community of interest" no longer exists, the relationship is no longer beneficial to either party. At the same time, the termination and renewal processes must be monitored to prevent abuses.

The committee's report also outlined four general policy objectives essential to developing regulatory policy that balances the interests of franchisors and franchisees in the critical areas of franchise termination, renewal, and transfer. These objectives are:

1. To monitor the process of termination and renewal to limit abuse and permit cancellation only where undertaken with good cause and in conformity with franchise agreements.
2. To provide franchisees with sufficient notice and appropriate opportunity to correct failures or deficiencies cited as cause for such action.
3. To facilitate termination or nonrenewal where cause exists and where deficiencies have not been corrected.
4. To provide assurance to franchisees that upon termination or cancellation there is opportunity to recover some portion of the value of their investment in a franchise.

In conclusion, the report of the committee stated:

> Congress must initiate the process of establishing an appropriate balance in franchise regulation between the interests of franchisors and franchisees, and between the interests and capabilities of federal and state regulatory agencies. Essential to such an effort is the development of national standards governing conduct of franchise relations. In much the same manner as federal requirements have established minimum national standards for franchise disclosure, federal legislation is required to provide minimum standards of protection for the legal rights and financial interests of franchisees.

National standards of franchise conduct should be adequate in themselves not only to provide sufficient protection to franchisees but also to encourage uniformity in franchise arrangements nationally. At the same time, states should retain the right to provide additional protections and to continue existing or expanded enforcement efforts. Such standards should incorporate:

- the extension of generally recognized principles of good faith and fair dealing to all aspects of the negotiation, performance, and severance of franchise agreements
- the creation of a federal right of action for franchisees and potential franchisees who sustain loss or injury as a result of violations of federal requirements relating to franchise sales and disclosure and to the conduct of franchise relationships
- the restriction of the use of procedural devices in franchise agreements that seek to shield franchisors from legal liability or inhibit the exercise of legal rights by franchisees
- the development of federal standards of good cause relating to terminations, renewals, and transfers of franchises. Such standards should include minimum requirements for notice and provide opportunity to correct deficiencies prior to action.
- the establishment of federal guidelines defining the circumstances and financial criteria under which a franchisee may

be entitled to remuneration by a franchisor for a portion of the value of a franchise upon termination or cancellation

These recommendations are *proposed* actions and deal with situations where the franchisee-franchisor relationship is in serious jeopardy. Only a small percentage of franchisees are not satisfied, and many problems can be solved prior to reaching a critical stage. The symbiotic relationship between the franchisor and the franchisee is the primary basis for the success of franchising.

Restaurant Study 14

Blimpie International

In 1964, Tony Conza and two high-school buddies created and sold their first Blimpie submarine sandwich behind a store counter in Hoboken, New Jersey. "Within the first year, we sold our first franchise for $600," said Conza. "After that, business kept getting better and better."

Currently, the company has more than 2,000 franchised Blimpie Subs & Salads outlets in 46 states, and 12 foreign countries. These outlets are traditional restaurants as well as new concept locations, such as college campuses, convenience stores, and sporting arenas.

Consumer demand for nutritious and convenient meals is on the rise, with no sign of abatement. Blimpie believes their menu is more timely now than ever before in their 34-year history. And they have the statistics to prove it!

- Restaurants rank as the leading suppliers of prepared food consumed in the household, with $33.6 billion in sales, and growing at a rate of 4.6 percent annually. The sandwich segment alone accounted for an estimated $11 billion in this category. *(Source: Technomic, Inc.)*
- Branded fast foods became the second leading revenue stream at convenience stores, accounting for 9 percent of total convenience store sales. *(Source: NAIC State of the Industry Report)*

Blimpie is positioned to capitalize on all of the above-mentioned consumer trends and opportunities. Their menu offerings accommodate Americans' growing preference for fresh, nutritious food over fried food. Their freshly made sub sandwiches and salads provide quick, delicious home meals. And with more than 500 outlets in convenience stores and a successful new-concept strategy, the stage is set to expand their role in this important category. Most people have a desire to eat right, they just don't have the time to search out places that offer good, healthy food. Blimpie reflects exactly what consumers want. Today, Blimpie outlets can be found in new-concept outlets such as convenience stores, mini-marts, colleges, schools, hospitals, sports arenas, mass merchandisers, and institutional food-service operations.

In 1996, Blimpie Subs and Salads reached a milestone—1,500 stores—a number that triggered an increased national advertising contribution. Consequently, the company will aggressively increase its advertising this year as a way to continue building the brand worldwide.

By hiring *kirshenbaum bond & partners,* New York, a highly regarded, innovative advertising agency, Blimpie International boasts an integrated marketing initiative designed to convey the brand's distinct advantage over its competition—Blimpie makes the best sandwiches, because of the uncompromised commitment to quality and passion for excellence.

For the chain's first nationwide campaign in 1996, the agency persuaded Conza, Chairman and CEO of Blimpie, to star in TV and radio spots to express his passion for the sandwich, in witty scenarios, under the slogan: Blimpie. It's a Beautiful Thing. The campaign which continues today, includes print and broadcast ads, point-of-purchase materials, and signage, and is designed to make customers think of Blimpie first whenever they crave a sandwich.

Blimpie International (BI) continually searches for ways to improve the energy of BI people. All Blimpie Subs & Salads outlets are owned and operated by an extended family of franchisees. In fiscal year 1997, the family grew significantly, with 426 new outlets opened.

BI firmly believes in giving back to the communities in which it operates. As a proud supporter of the Boys & Girls Clubs of America (BCGA), the company's involvement ranges from donating food to children's events at the local level, to raising $100,000 as part of its Summer Combo Meal promotion to a commitment of $500,000 from Blimpie International to the BCGA by the end of the year 2000. CEO Conza serves on the club's national board of governors.

The Blimpie Subs & Salads brand is taking on new challenges as the company focuses on the future. Embarking on plans to take the Blimpie sub overseas, Blimpie International has forged partnerships with sub-franchisors in Europe, Asia, and Latin America. Blimpie recently expanded to Cyprus, South Africa, and Jordan. Other international expansion is planned throughout 1998 in addition to the current international locations.

In October of 1997, Blimpie International announced a comprehensive strategy for continued growth that includes franchising a family of brands. BI will be the majority owner in a new company, Maui Tacos International Inc., and plans to franchise this "Maui-Mex" concept called Maui Tacos. To further address current eating trends, Blimpie International has also launched a new concept called Pasta Central, which will offer an assortment of pasta meals in restaurants and as home-cooked meal replacement for consumers on the go. A third concept has been introduced called Smoothie Island. Smoothie Island is a concept that offers fruit-based beverages that can be supplemented with a variety of vitamins and minerals. To provide additional franchise services, BI Concept Systems, Inc. has been incorporated from the former equipment

(a)

(b)

An exterior view of a Blimpie International restaurant (courtesy Blimpie International)

department to service independent entities outside the BI family of brands. This subsidiary of Blimpie International was created to respond to the need in the industry for efficient, quality equipment systems. BI Concept Systems provides these turnkey systems to other companies in the same timely and cost-efficient way that has made the Blimpie franchise system such a success. The company is set up to support a restaurant opening from inception to completion, including preliminary site work, postopening warranty issues, and service calls.

CHAPTER *10*

International Franchising

Franchising in the international market is expanding rapidly; in particular, restaurant franchises have seen a tremendous increase in recent years. American restaurant franchises are now prevalent in almost every corner of the world. The United States, pioneer in the field of franchising, continues to be the worldwide leader in restaurant franchising. According to National Restaurant Association researchers, over 160 U.S. foodservice companies are operating internationally. The quick-service segment of the restaurant industry is the fastest-growing segment in foreign markets. It was reported that in 1994, nine out of ten international companies, in terms of sales, were in the quick-service segment. It was reported in *Restaurants USA* (Nov. 1995) that in 1994, McDonald's opened more than 1,200 restaurants worldwide; almost two-thirds of those restaurants were outside the United States. McDonald's boasts of cutting the ribbon on a new restaurant somewhere in the world every four hours, on an average. They seek to dominate before the competition even establishes a foothold. They claim that most of the growth still comes from existing markets, particularly the Big Six—Japan, Canada, Germany, England, Australia, and France—that have the most restaurants. It was also reported that Extra Value Meals are powering McDonald's sales in the above-mentioned and other established markets, some of which sell a higher proportion of Value Meals than is sold in the United States.

Some franchise corporations are planning to build more units overseas when compared to the United States. Business-format franchising is the most common type of franchising currently being used for expansion in the international restaurant market. It can rightly be said that one of the most rapid and successful expansions of U.S. business has been in restaurant franchising. For this reason, the discussion in this chapter is heavily slanted toward the U.S. franchise restaurant industry, although most factors discussed have universal applicability. Many franchisors who are entering foreign markets have been very successful in the United States and have the necessary expertise to succeed in international franchising. Although some attribute expanding international markets to domestic saturation, scores of other reasons make foreign markets appealing to U.S. franchise restaurants. One aspect that has definitely promoted American restaurant franchises overseas is the standard quality of the products and services they offer. This serves as a major selling point for many of the franchised products and services.

Notable trends that favor the increase of international franchising in foreign markets include (a) increased educational status of the local population, (b) technological advancement facilitating travel, intercultural cooperation, and instant dissemination of information, (c) exposure to different foods; the willingness of younger generation to try new products and unconventional types of foods, (d) rapid development of rural areas, construction of highways, improved transportation methods, and overall industrial development, (e) improved economies and increased disposable family income, (f) the increased numbers of women in the workforce and of two-income families, (g) the increased significance of convenience as a result of one or more factors mentioned above, and (h) the popularity of take-out or home-delivered meals. The most important factors are outlined and discussed below.

Factors Related to International Expansion

Expanded Market

International markets provide new dimensions for the expansion of restaurant franchises. Increasing population, which is manifold for some countries, and the increase in available disposable income has created an expanded market. Based on the size and increase in the population, the combined market size and potential demand for franchise restaurants is much greater in some other countries than in the United States. Demographics of countries such as China, Korea, Malaysia, India, and Indonesia are changing rapidly. China and Vietnam are expected to adapt restaurant franchising, which is expected to develop very rapidly.

The financial status of foreign populations has changed for the better. Because of the increased export of natural products and finished goods, many people have more economic resources. It is estimated that the U.S. population will grow from 250 million to 283 million between 1990 and 2010. However, during the same period the world's population is expected to grow from 5.3 billion to 7.2 billion. In many countries, untapped population growth and industrial expansion are taking place simultaneously. This prosperity has created large numbers of people who want to avail themselves of restaurants.

Economic and Demographic Trends

Economic and demographic trends that favor the increase of restaurant franchising in foreign markets include:

1. Increased educational levels of local populations around the world are promoting an awareness of different cultures, foods, tradition, geographic destinations, and so on. This increase in educational development has broadened technological skills as well.
2. Technological advancement has led to the widespread use of computers, electronic equipment, and other facilities. This advancement has also facilitated travel and tourism, bringing better awareness of different countries as well as exposure to different types of franchise opportunities.
3. There is an increased willingness of the younger generation to try new products and unconventional types of foods. Teenagers are more apt to try different types of foods of which they become aware due to technological developments.
4. Rapid development of rural areas and the concentration of population in urban and industrial areas have led to an increase in the number of people eating away from home. This has indirectly led to the development of franchised restaurants.
5. With the economic growth comes an increase in the available disposable income of the families. This increased spending power has helped in the growth of franchised restaurants.
6. With all the industrial development occurring around the world, women have entered the workforce in a considerably large numbers. The increase in two-income families has also led to eating away from home either during lunchtime or in the evening for dinner.
7. There is an increased emphasis on convenience due to many of the factors listed above. Franchised restaurants provide the convenience for which modern consumers are looking. Technological advancements have brought computers and videos in almost every home, which has increased the preoccupation of families after work; this again promotes the eating of foods

prepared and/or delivered from outside of the home. This is also the reason for the popularity of take-out and home-delivered menu items.

These factors were primarily responsible for the development of franchises in the United States and now the same phenomenon is being witnessed in the worldwide arena.

Increased Travel and Tourism

Increased travel and tourism for business or pleasure have positively exposed the successful and rapidly growing U.S. restaurant franchise industry to visitors from other countries. This has not only introduced travelers to the quick-service industry and its food products but also has encouraged businesspeople from other countries to develop an interest in investing in restaurant franchises.

Quality of Products and Services

American restaurant franchises are known for the quality of the products and services they provide, which are regarded highly in many countries and serve as a selling point for franchised products. In fact, many feel it is safe to eat at an American franchise restaurant in places where one is not sure of the safety standards or the quality of food in other establishments. The quality of food served by U.S. franchised restaurants is standardized and definitely superior when compared to similar food products served in some foreign countries, thus increasing the popularity of American-style foods.

The demand for food products offered by American franchise restaurants in many countries may be greater than we can imagine, although this may not be true for all countries. There are many U.S. franchised restaurants overseas whose sales volume is greater than any unit in the United States. Some of the largest sales-volume-producing restaurants of the U.S. quick-service restaurant franchises are located overseas. This will increase with the growth of international franchises. A good example is the demand for products like hamburgers, fried chicken, pizza, Pepsi®, and Coca-Cola® overseas. Many products were relatively unknown outside of the United States. An example is that of Japan, where until recently pepperoni was an unknown commodity. However, now it is one of the most popular items. Acceptance of any food item is greatly dependent on taste development over a considerably long time with frequent exposures. Once the taste is developed and the product is accepted, then it may become popular. Many menu items that are popular in the United States may take some time to be accepted overseas. There will always be some population segments in the world where certain products may not be readily or even eventually accepted. Thus, restaurant franchisors must plan their strategies carefully and target sections of the world where there is a good potential for achieving acceptance, familiarity, or popularity of their foods.

American restaurant franchises have standards and symbols that are highly favored by many people, nationally and internationally. In some countries, eating at franchise restaurants of American origin is considered a status symbol. In Taiwan, for example, the younger generation considers eating at an American franchise restaurant a special social occasion. Also, burgers and fries are seen as a Western experience and have become symbols of American culture. There are expectations of standard products, fast service, lighted and air-conditioned dining areas, and clean rest rooms, which are mostly associated with franchise restaurants. The standards of quality demonstrated by the adherence to quality, service, value, and cleanliness (QSVC) are a primary factor in their acceptability. These standards of quality are widely publicized in many countries. In Taiwan, QSVC is promoted and explained to customers. Also, as a confidence-building effort, McDonald's solicits questions from consumers and answers are posted in big letters on a bulletin board near the counters in the restaurant. Quality and safety standards set by American franchises make their products and services consistent in quality and, above all, safe.

There is a willingness among individuals in the younger generation to try new and unconventional food products. Foods offered by American franchises are more acceptable now than ever before, especially by younger people in almost all parts of the world. Nutritional concerns related to fast foods in the United States are not necessarily considered valid in some countries, where undernourishment rather than overnourishment is the major concern.

This expansion of U.S. franchises is not without criticism. It is claimed by some that this expansion is exploiting the indigenous cuisines, local businesses, and national cultures. However, this criticism is somewhat weak, considering the overwhelming popularity of the U.S. franchise restaurant's menu items.

Technological Advancement

Technological advances have led to sophisticated business controls and the worldwide adoption of related management techniques, making it easier to implement franchise systems internationally. Technological advances have led also to increased movement of populations to urban and industrial areas, resulting in increased demand for convenience food. Computers, videos, and other electronic items are creating needs that are unparalleled in the history of humankind.

Cyberspace technology is adding an interactive computer environment that facilitates many activities. McDonald's has developed McFamily®, a friendly, on-line, interactive computer environment that parents and children can explore together. Its subjects include fun and information pertaining to parenting. It is considered a great way to communicate and build strong relationships with their customers. The concepts of eating away from home, at drive-ins, or ordering home-delivered meals are gaining popularity. Franchise restaurants are already designed to serve these concepts and therefore have ready applicability in many parts of the world.

Business Management

Many countries have already borrowed American ways of conducting business, in which franchising plays an important role. American entrepreneurship and the role of the multi-billion-dollar quick-service industry in the U.S. economy have attracted foreign entrepreneurs to seek replications in their own countries. Some significant factors to consider when contemplating international expansion include the relatively easy availability and lower costs involved in securing human resources, ingredients, and products in a foreign market. The sales volume required to break even or to offset possible losses may be smaller than expected. Expenses, such as for advertising and training, may also be relatively less, thereby increasing profitability.

A word of caution is necessary at this point: the above mentioned positive factors may be reversed in some countries. The extraordinarily high costs of start-up, occupancy, labor, and food in Tokyo are blamed for delayed or nonexistent profits. It may be easy to find employees in Indonesia or the Philippines, but the available disposable income may be so low as to make the products unaffordable by the average person. Thus, there is a need to carefully balance positive and negative factors. Although finding this balance may be difficult, it is not impossible, and it is worth the effort. Products offered by restaurant franchises and their very functionality have been successfully tried and tested in the United States to such an extent that it makes sense to adapt them for use elsewhere.

With the advancement of technology, educational standards, and economic conditions in many countries, a growing number of management-oriented entrepreneurs is available, with enormous potential to be successful in the operation and management of franchised restaurants. Multimedia teaching tools, publications, and computerized programs can be effectively used in management training.

Trade and Monetary Balance

International trade and monetary balance has changed significantly in recent years. These changes have made investment of American dollars in foreign markets, and in some cases the other way around, very favorable. Fluctuating currency values remain a major concern of foreign investors and play a role in investment decisions at home and abroad. For example, financial crises faced by many Far East countries have affected overseas expansion or market entry by many U.S. franchisors.

Political Climate

More openness in the Eastern European bloc countries has created a curiosity about and a tremendous potential for the growth of franchise restaurants there. Such political changes, if favorable and of long-term duration, can provide opportunities unknown in the past. The chances of a franchise flourishing in places where none now exists are great and should be developed to the maximum by all franchisors considering international expansion.

When McDonald's became the first fast-food chain to open a restaurant in a former Soviet bloc country, people stood in line for three hours to sample American burgers! Similar scenes were later seen in Russia. Pizza Hut and Baskin-Robbins® have entered in new markets in addition to McDonald's and KFC. The political climate has changed considerably in many countries, creating a favorable environment for U.S. business ventures. Political stability in a country makes the atmosphere conducive for business. Major changes in the world scene are emerging from the changed political system in Eastern Europe and the unification of the European Economic Community (EEC).

The political and geographical changes will result in the free movement of goods, services, people, and capital resources within the community; centralization of purchasing and distribution functions; pooling of human resources; and uniform codes, specifications, and regulations. All of these changes will have a profound impact on franchise businesses being considered or planned for the future.

POINTS TO CONSIDER IN INTERNATIONAL FRANCHISING

The significance and potential of international franchising cannot be underestimated. However, there are points to consider before the decision is made to enter foreign markets. These factors are discussed below, illustrated with examples of existing franchises in selected parts of the world.

Political Environment and Legal Considerations

Political stability and legal restrictions should be thoroughly investigated before planning to enter any foreign market. A government that favors franchising or a particular franchise may be replaced by one that is hostile. It may be difficult to gauge stability, but past history and the political environment may give a good indication. In addition to adverse political situations, some countries have legal restrictions that may make it difficult for U.S. franchises to function.

Foreign governments may have regulations that make it difficult for any franchise to be developed in a particular city or region. For example, the French government has created a historical classification aimed largely at protecting the traditional decor and atmosphere of French restaurants. This restricts the use of bright lights and the atmosphere normally prevalent in fast-food restaurants. The terms and conditions of contracts must therefore be written in a way to allow room for exit or adaptation to unfavorable situations.

Associated with the political environment are the monetary restrictions a country may have. If complex bureaucracy is involved or if it is difficult to get money out of the country, franchisors may be at a disadvantage. A fluctuating currency may also have an impact on the profitability of the operation. World currencies that exert considerable influence on franchising include the German mark, the French franc, the British pound sterling, and the Japanese yen. Fluctuations in these currencies influences profits and losses incurred on foreign investment. McDonald's lessens short-term cash exposures by primarily purchasing goods and services in local currencies, financing in local currencies, and hedging foreign-dominated cash flows.

On the other hand, there are countries that offer incentives for foreign investors and provide special considerations. Such information may be obtained from trade missions within embassies or offices of the counselors of the respective countries. Franchisors should also be informed about local taxes and tariffs associated with conducting business in a particular country.

Import laws and regulations also play an important role for franchise restaurants. Many countries have regulations that restrict the import of any product(s) that can be secured domestically or those that contain certain ingredients or additives. When International Dairy Queen® wanted to establish a restaurant in Korea, the government provided the company with a list of items that could be imported. However, when it tried to import toppings for sundaes, the company found that it could not import anything that had preservatives. Such problems are multiplied when perishable items, such as meats, are concerned. Finding alternatives or coming up with modifications become tough questions and pose difficult problems.

Because restaurants and real estate are linked, real estate laws and regulations have a significant impact on franchises. In several countries, there is no protection or compensation against real estate loss that results from private or governmental decisions. Also, sometimes no explanation or reasons are given for the action. This uncertainty is a leading cause of hesitation for many franchisors. Also, lease agreements vary from country to country. Franchise renewals become uncertain under such circumstances. In Vietnam there are no rights of ownership for properties, as all real estate belongs to the government. Such laws are prevalent in many countries. In addition, there may be problems and differences stemming from local and city governments that pose a threat to the investment made on franchise restaurants. KFC in India was reported to face problems with local health officials and city governments that imposed restrictions on and even forced the closure of restaurant units. According to the newspaper reports one restaurant had two flies, which led to its closure. However, other units were closed at the same time, alleging that the products contained cancer-causing chemicals. These actions are a result of veiled political intentions for which unreasonable methods are sometimes used.

Some countries have quotas that must be filled for foreign investment, which should be taken into consideration. Additional charges may be levied for the processing of papers, and some countries may have restrictions or embargoes on specific product(s) from other countries. All paperwork involved should be thoroughly investigated. Developing countries with stable political and economic situations are good potential candidates for American restaurant franchises. Political and economic stability is important because fluctuations in exchange rates as well as the possibility of money transfers have a considerable impact on the success of any franchise overseas. Sometimes creative solutions and alternatives have to be designed to overcome such situations.

Politics plays an important role in international franchising, and a phenomenon known as *guilt by association* must be considered. No matter what the economic significance of a franchise is to a particular country, American franchises will always be identified with the United States. There may be occasions when these franchises become targets of physical or nonphysical attacks by a government or the public in general. Such situations arose for American franchises in China, Beirut, and Nicaragua, where businesses had to be closed. In the past, Kentucky Fried Chicken had to temporarily shut its two Beijing locations, one of which is in Tiananmen Square, out of concern for employee safety. McDonald's in Taiwan had a bomb scare and scores of restaurants faced problems during periods of political rifts. Insurance against such risks is hard to obtain in foreign countries.

Language, Culture, and Traditions

Language plays an important role in the successful operation of a franchise system in foreign countries. A good working knowledge of the language of the country is essential, primarily for three purposes: (1) for effective communication with franchisees and for the complete delivery of the philosophy, strategy, and functioning of the franchise system, (2) for developing and implementing a successful training program for employees and management personnel, and (3) for providing details of operations and for the development of operation manual(s) for franchisees.

The language barrier sometimes is a big deterrent to expansion. Sometimes communication effectiveness is lost in translation. Even the pronunciation of certain trade names become difficult or meaningless. For example, there is no *P* sound in Arabic, so Pepsi becomes "bebsi" and Popeye's is pronounced "bobeyes." This is not as embarrassing as when the meaning of words or sentences is affected. It was reported that, when Pepsi translated the slogan "Come alive with Pepsi" into German, it became "Come out of the grave with Pepsi." In Chinese the same slogan became "Pepsi brings your ancestors back from the dead." When KFC translated the slogan "Finger lickin' good" into Chinese, it became "Eat your fingers off." This problem is not restricted to foreign languages only, as even in English-speaking countries words can have different meanings. However, countries where English is spoken and understood, such as the United Kingdom, Australia, Hong Kong, Malaysia, India, and Singapore, present no problems.

Language should not be an absolute deterrent, however, as in countries like Japan and Taiwan, U.S. franchises have functioned successfully. For example, there are McDonald's Hamburger universities located outside of the United States. One opened in Tokyo in 1971; Munich followed in 1975 and London in 1982. The academic curriculum at the international Hamburger University locations is the same as used in Oak Brook, but adapted for each country's individual needs and cultural differences.

Products that are acceptable and extremely popular in the United States may not be acceptable in all countries. When Domino's Pizza went to Japan in 1985, the company discovered that there was no Japanese word for pepperoni and that the Japanese had no taste for pepperoni. Domino's imported pepperoni until the Japanese were trained to make this product. After more than ten years, pepperoni pizza is the number-one-selling pie in Domino's Japanese units.

In many countries, it is difficult to import products that are available in the country. In South Korea, pepperoni cannot be imported because the government refuses to allow pork to be imported. Likewise, in many countries, including the EEC, it is hard to import cereals due to their high sugar content and the use of artificial additives. Some elements of franchising that may be considered minor present significant stumbling blocks in other countries. For example, a trade name and trademark may have a meaning in another language that is not considered decent or acceptable.

Culture and food habits have an enormous impact on food preferences and selection. Certain countries find food items such as pork and pork products, beef, and alcoholic beverages unacceptable. Other animal products or combinations of certain ingredients may not be accepted in other countries. An example is seen in Malaysia, where American franchises have signs on their menus and counters showing that all meat products are *Halal*—an indication that Islamic slaughter techniques were used, an important consideration in countries in that region and in the Middle East. The prohibition of pork products poses a problem for the selection of breakfast items by franchises. Thus the popularity of breakfast in the United States does not necessarily translate into profitability in other countries. Meal timings also vary considerably, and eating breakfast out may not be a popular idea in many countries. Sometimes names contradict with the beliefs of certain population areas. When AFC (America's Favorite Chicken Company) wanted to register their Churchs® trademark in countries like Taiwan and Saudi Arabia, the government refused to accept it because the name carries a religious connotation. Thus, in those countries, Churchs uses the name Texas Chicken®.

Eating styles may be distinctly different. Vast numbers of people may not like the idea of holding sandwiches in hand (without knife and fork) and biting with their mouth wide open! Prospective franchisors should be aware of and study all such food habits and cultural factors, which will have a tremendous impact on the success of a restaurant business.

Menu Items and Service

Not only the menu items selected but also variations of each item must be considered. Tastes vary from population to population. For example, fried chicken served by Kentucky Fried Chicken in Taiwan and Malaysia is considerably spicier than in the United States. Chili sauce is provided with chicken in Mexico and in South American countries. These types of modifications are designed to make the product suitable for the taste buds of the local population (see Figure 10.1).

Menu selection is important, because what is popular in the United States may not be popular elsewhere. Some items, such as hot dogs, pretzels, and baked potatoes, are unknown to many

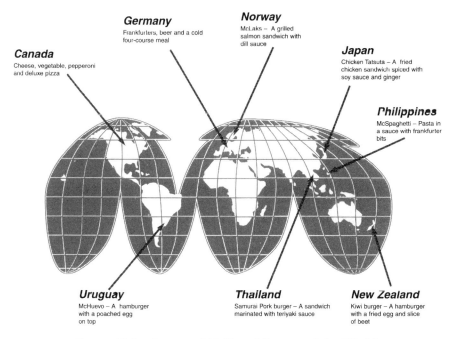

FIGURE 10.1 A taste of McDonald's around the World

people overseas. In fact, in their languages, Portugal and other European countries do not even have a word for popcorn!

Fried chicken and burgers are the most popular items in many countries. It is interesting to note that, depending on the country, franchise restaurants try to attract both the chicken- and the hamburger-eating segment of the population. Kentucky Fried Chicken sells Colonel Burgers and McDonald's sells fried chicken in many of the Pacific Rim countries. Salads are not as popular in many countries as in the United States. Eating habits play a very important role in the popularity of menu items. In the Pacific Rim area, where fresh fish and fish products are readily available and popular, frozen fish products may not be well received. Some modifications and additions to the menu may be necessary. McDonald's in Malaysia, for example, uses a popular local type of dried meat in the burger and includes sugar cane juice on the list of beverages.

Tropical fruit juices and toppings other than those offered in the United States must be considered for inclusion on the menu.

In India, McDonald's offers a vegetarian hamburger, as most of the population do not eat beef. In many countries, chicken is more popular than beef. In Japan, rice burgers are popular. In Taiwan, McDonald's serves hot soups in drinking cups and advertised a soup bowl set that could be purchased in conjunction with menu items, for promotional purposes. In addition to soup, McDonald's has teriyaki burgers on their menu. Texas Chicken has a shrimp sandwich, chicken soup, corn chowder, buttermilk pie, and sweet corn nuggets as special menu items. Pizza Hut in Taiwan offers a wide variety of pizza selections, including seafood, Italian, surf and turf, tuna, vegetarian, smokehouse chicken, mushroom, Mexican, and shrimp delight. A fresh salad bar, nicely displayed, is an added attraction at this Pizza Hut restaurant.

An important factor in food popularity is the taste development for a particular product. Hamburger is popular in the United States, but in many countries people may have no idea of its taste. Chicken in Middle Eastern countries is roasted, it is used in conjunction with curries in Asian and Far Eastern countries. Fried and batter-dipped chicken may need time for taste development. In the United States, the gradual development of a taste for pizza illustrates the point that time is required for full acceptance. The same was true in Japan. Thus, restaurant franchisors should target sections of the world in which there is a good potential for achieving acceptance or popularity of their product in a reasonable period of time. In many countries, it is common to see the younger generation at fast-food restaurants; however, this trend is changing rapidly and many senior citizens are patronizing these restaurants, particularly when they go out with young children.

Internationally, there are notable differences in services. In countries where eating at a dining table is customary, take-out and delivery systems are not well accepted. In some countries, chinaware and silverware is used rather than paper or polystyrene products. Also, in some countries, trays and soiled dishes are left on the table and later cleared by the waitstaff even in quick-service restaurants. Elsewhere, consumers prefer U.S. fast-food restaurants because of cleanliness, comfort, such as air-conditioning, and because the novelty of the American concept itself appeals to many adolescent consumers. Business ethics, etiquette, and customs vary from country to country, which should be taken into account. From menu selection to service, all of these aspects are important to the success of a franchise.

Cultural differences have an impact on what can be marketed and what type of service is preferred. For example, in many countries of the Middle East and Far East, breakfast is not eaten away from home and does not consist of items that are popular in the United States. Thus, it is not proper to offer breakfast menu items or to open the restaurant during breakfast time. In Thailand, for example, people prefer soup items in the morning, and in other Asian countries breakfast consists of very different menu items.

The type of service preferred is also affected by the cultural norms of a country. In Saudi Arabia, women do not freely eat with men in restaurants when they go out, which is very seldom. In restaurants, family dining areas must be separate from dining areas for single men. In many countries, the concept of standing in line to order is not expected. It is not uncommon to see large crowds standing in front of a restaurant in Taipei waiting to purchase foods from local restaurants. That is one of the reasons why in U.S. franchised restaurants in Taipei a waitress at the door fills in each order on a menu card and hands it to the customer, who then waits in line. This also helps in saving time for menu selection by customers who are mostly unfamiliar with the items served by the restaurants. Also, in some countries, women and men do not stand in the same line.

Delivery service also faces critical problems in many countries and it is not as easy as in the United States. Often it is hard to locate addresses and roads are complicated. Also, in Asian countries and former Soviet bloc countries, telephones are not very common in households. For example, in Poland, only 30 percent of homes have telephones, and in some cities in India the percentage is much lower. Therefore, in such countries typical delivery service restaurants like Domino's are forced to operate restaurants with seating. Furthermore, traffic in Thailand and other countries is so congested that it is almost impossible to deliver food on time. Under these conditions food is delivered on scooters or motor bikes, which can pass between cars and through alleys easily. A insulated box on the back of motor bikes keeps pizza and other food items warm.

Because real estate is expensive and open spaces are hard to find, in many cities franchise restaurants are located in the midst of buildings or have several stories. McDonald's in Taipei, for example, is on four levels, with a children's play area located on the fourth floor. It is also due to the lack of space that drive-throughs are not commonly seen in developing countries, where gas expenses prohibit extensive use of cars.

Careful site selection is important, and prospective franchisors should research and study all relevant factors. For example, highway travel may not be as popular in some countries as it is in the United States. On the other hand, train and bus travel may be more popular. Walk-in traffic is common in many city centers due to the lack of parking facilities. Aspects such as the architectural design and colors used in a restaurant are very important. Height and lighting restrictions pose considerable problems for franchise restaurants in Paris. Similar problems are faced when using signage. Due to the lack of space, it is not possible to have large billboards or signs, as seen in the United States. Also, free-standing restaurants are uncommon in large city centers, primarily due to lack of business space.

In heavily populated urban regions like Tokyo, there may be a heavy demand for food at peak times, necessitating extremely fast service.

Demographic and Economic Data

An accurate demographic study of the target population is essential before and during the operation of franchised restaurants on foreign soil, and should be undertaken as a part of the feasibility study. Age, sex, and available disposable income of target families should be evaluated, with special emphasis on the potential for changes in the near future. This type of study helps in planning and managing the operation. Demographic changes may be more distinct and rapid in other countries than is expected in the United States.

Consideration must be given to the developmental stage of the target country. Some countries acquire wealth in a short time from natural resources and have ambitious plans for development. Educational status and working conditions may change with a growing number of educational institutions. Large segments of the population may migrate or become expatriate, particularly in the Middle East and some Far Eastern countries. All these changes make the need for restaurant facilities planning significant and essential.

American military bases and centers abroad are good locations for franchised restaurants. In short, there are more mouths to be fed and, consequently, more opportunities for foodservice establishments. Many restaurants are located in industrial centers or malls and shopping complexes. In many countries where franchise restaurants have been started, the business is flourishing. In fact, some franchise restaurant units outsell all other restaurants in their respective countries and regions. Careful study of the demographics and the economic status of a country is vital for the planning and future success of American franchises.

Menu prices and cost control methods have to be adapted to particular situations. The value of the dollar and the cost of ingredients can make the overall cost of menu items too high and out of reach of the average person.

Availability of Resources

An important element in international franchising is the availability of resources, as is the possibility of conformation to the standards set by the franchisors. For example, chicken or tomatoes (per the specifications of the franchisors) may not be locally available or, if available, may not be in sufficient quantities to keep pace with demand. McDonald's, for the first time, took responsibility for meat, potato, and other processing operations, rather than giving it to vendors when it opened its unit in Greater Moscow. This type of arrangement may be necessary in some countries in order for franchising to be successful.

Compounded with these factors is the fact that indigenous products and ingredients may not function as expected when combined, adversely affecting the quality of the finished product. For example, if fresh biscuits are to be prepared, flour and other ingredients (obtained from local sources), when mixed, may not result in a product of the desired quality. Seasonal variations, varietal characteristics (particularly of vegetables), and the availability of all items should be considered.

Equipment availability and performance may vary, which can have a distinct impact on the quality of the product. In certain countries, power and water supplies may be limited or regulated, affecting the quality of product and services. Types of storage facilities and temperature may dictate the extent to which products can be stored. In short, the potential for conformation to the standards and specifications of a franchisor should be carefully assessed. Due to lack of space, both the back and front parts of the restaurant may have to be remodeled. Optimal use of space restricts the use of large equipment for storage and production. This also has an impact on the amount and kind of inventory that can be maintained. Also, food and traffic flows within the kitchen areas have to be modified.

For preserving proprietary rights of trade formulations and specifications, reliable supply sources should be used. In some countries, a limited number of suppliers may be serving several restaurants at a time. In order to maintain the competitive edge, some franchisors may need to consider developing their own procuring and processing facilities.

In order to overcome the many differences and problems outlined in this chapter, U.S. franchises often prefer going with either joint ventures or master franchising. Local partners are familiar with the local situation and can contribute alternative solutions. Also, they are more familiar with the bureaucracy and are able to handle it better than a total foreigner in many countries. It becomes evident to many U.S. businesses that a great deal of patience and tolerance is needed to be successful in many foreign markets. Flexibility, modifications, and creative adjustments are necessary.

Technology Transfer

The growth of franchise restaurants in the international arena is contributing to the import of technology and knowledge of franchising systems overseas. The knowledge of franchising itself not only provides a way of conducting business but also serves as an excellent example of the free enterprise system to emerging democracies around the world. The use of the system stimulates the economy and the development of infrastructure within a country. Also, the concept of franchising itself is an attraction because it demonstrates a chain reaction where a restaurant can help the national economy. These restaurants buy food products from local sources and help develop agriculture, food processing, and commercial fishing. In addition, franchises create job opportunities and help local communities. Also, the training provided is valuable. In other words, franchising, in its own way, is helping to develop new market economies. This in turn helps in the exchange of technology and information.

All the above-mentioned factors force franchisors to reevaluate their strategy, operational parameters, and way of conducting business utilizing the technology available to them. The transfer of technology takes place in both directions between the United States and other countries. Knowledge gained from experience overseas helps in remodeling or redefining needs, which translates into benefits to franchisors, who can use the modified technology and processes in franchises in the United States.

Franchisors in the United States have been strict in adhering to menus and service concepts. The international experience has made them realize the significance and usefulness of menu modification. Many of the lessons learned are useful in allowing modifications of menu items and providing flexibility in operations to franchisees. New menu items used in other countries may well prove of use in the United States. Also, as the demographics change in America, resulting in increased cultural diversity, there will be increased need for international menu items or a combination of international and domestic items.

Franchising facilitates the transfer of technology in more than one way. In the case of franchised restaurants, there are both tangible and intangible aspects. The first one deals with products while the latter deals with services. Because restaurants involve both products and services, technology transfer takes place in both. The transfer of tangible aspect concerns work ethics and habits that result from training and working in a franchise operation.

Tangible aspects in technology transfer related to franchise restaurants include training, operational methods, operations manuals, equipment technology, process information, and selected service parameters. The intangible aspects include the conduct of services, which cannot be measured. Both of these aspects transfer with franchising. Even the knowledge of how to conduct business using franchising is an important technology transfer component. Franchising has created a spirit of entrepreneurship that has resulted in a rapid growth of franchising as a method of doing business. This entrepreneurship has also generated interest among prospective businessmen around the world. With the advancement of technology, educational standards, and economical conditions in many countries, a growing number of management-oriented entrepreneurs who have the potential to be successful in franchised operations is available. Also, multimedia teaching tools, publications, computer programs, and access to the Internet can be effectively used for management training purposes.

It is a known fact that McDonald's has trained indirectly millions of high school students in work ethics, customer interaction, the value of service, and how to be self-reliant by working in their restaurants. This has made an enormous impact on a generation. Similar contributions are made by other types of franchises. This benefit becomes available to other countries as franchises move abroad. Their impact in different cultural situations will itself be an interesting study, the results of which will be evident a few years after the franchise system enters a host country.

Points to Consider in International Franchising

1. Include local businesspeople in decision making, planning, and the operation of a franchise.
2. Base the concept on the local population's preferences, food habits, and cultural features.
3. Be sensitive to the needs of the local population and be respectful of their beliefs and practices, both political or religious.
4. Be patient and tolerant and follow the legal steps with as little challenge to rules and regulations as possible.
5. Conduct a good environmental scan, taking into consideration social, political, and economic circumstances of the country.
6. Closely examine the foreign exchange situation and the amount of the money allowed to go out of the country.
7. Examine the per capita gross domestic product and the population of the country.
8. Hire experienced people, with good knowledge of local business, prior to and after entering a foreign market.
9. Carefully examine the demographic data of a country, with emphasis on the emerging population that can be classified in the middle income group and relatively young.

10. Select menu choices carefully and be flexible and prepared for menu adaptations and modifications.

Methods and Mode of Entry

Direct Franchising

Direct franchising, often referred to as *licensing,* allows a franchisee to set up a franchise in another country using the system trademark, products, and services, and to function as does a franchisee in the United States. The franchisor franchises directly to franchisees situated in the foreign country without the intervention of a third party. This type of licensing is used by some restaurant franchises. Licensees are trained by the franchisor and assisted in the start-up and operation of the business.

In general, some restaurant franchises have foreign subsidiaries that are operated by corporations within the United States. Others have franchisees who are granted franchises by the company, a subsidiary, or an affiliate. An affiliate can be a company in which the franchisor has some equity, normally less than 50 percent, and the remaining equity being owned by a resident national. Some countries make it mandatory that a resident national be involved in licensing.

The success of other international franchise companies, irrespective of the products or services they sell, in a foreign country should be studied carefully. This assessment provides an overall picture of the international business environment as well as a basis for comparison.

Direct franchising can take one of three forms:

1. *Direct Unit Franchising.* The franchisor grants a franchise to an individual or a group directly from the country of origin in the same manner as he would grant a franchise in their own country. Thus, there is no difference in granting a franchise domestically or internationally.

2. *Establishing a Branch.* The franchisor establishes a branch office in the host country. This acts as the franchisor for the purpose of granting franchises in that country or the region.
3. *Development Agreement.* The franchisor enters into a development agreement directly with a developer who is a resident of the host country. Under the agreement, the developer agrees to develop and own all of the franchise outlets in the region.

In one of our studies conducted to find situations under which direct franchising should be considered, the following factors were ranked as important by franchise executives. In descending order, they are:

1. effective enforcement of and compliance with company's standards
2. control of brand penetration and brand image
3. greater control over franchisee
4. size of prospected country (If relatively small, a limited number of stores can be opened.)
5. low resources commitment by franchisors
6. significant differences in language, cultural, and legal system
7. no requirements for a franchisor to establish a branch or subsidiary
8. local tax issues

The advantages of direct franchising to the franchisor include the avoidance of having to keep pace with the changing requirements of a country, maintaining total control of the franchise system and trademarks, and retaining control over franchise functions such as advertising and promotion. The disadvantages can be attributed to the distance involved in managing and controlling. It may be difficult for the franchisor to service the franchise properly as well as provide training to foreign franchisees from the franchisor's home location. Also, decision making on matters that require rapid action may be difficult. This is further complicated when cultural and political issues are involved. From a legal point of view, it may be difficult for a nonresident franchisor to take legal action against a resident franchisee when necessary, due to jurisdictional restrictions. All of these disadvantages may also prove expensive and cost-prohibitive.

Master Franchising

Master franchisees act like minifranchisors in other countries. The franchisor enters into a master franchise agreement directly with a subfranchisor, usually a resident foreign national, pursuant to which the subfranchisor develops and owns franchise outlets in addition to subfranchising outlets to subfranchisees in the foreign country. Master franchisees may be individuals, businesses, or conglomerate corporations that assume the rights and obligations to establish franchises throughout a particular country or region. Potential subfranchisees deal with the master franchisee in all transactions.

Master franchisees may open their own restaurant or grant franchises to others. They also collect fees and royalty payments, which in turn they pay to the franchisor. Master franchisees perform most of the functions that a franchisor normally does, for which they are paid part of the collected fees or royalty fees, per the agreement. The agreement contract between the franchisor and the master franchisee clearly describes what is expected of each party and sets a specific period within which the master franchisee must meet stated objectives. Master franchisees have the responsibility not only to sign up subfranchisees within specific geographic areas but also to provide them with the training and support that are normally provided directly by the franchisor. Because master franchisees act as representatives of the franchisor, it is imperative to select them carefully based on their potential to perform efficiently as well as their ability to select subfranchisees. There is a need for a strong relationship and mutual reliance between franchisor and master franchisee. Situations under which master franchising should be considered were listed as follows in our study, in descending order:

1. the ease to support the franchise system
2. relatively prompt development
3. franchisors not having to make a capital commitment
4. significant differences in language, cultural, and legal systems
5. franchisors allowing local bussinesspeople to run their business in a manner consistent with local business practices
6. local tax issues

Master franchising has advantages in that master franchisees have knowledge of the local culture, politics, economics, and market for subfranchisees. The know-how, motivation, and entrepreneurial skills of subfranchisees can lower operations costs and help in making faster decisions based on local conditions. The disadvantages of master franchising include the difficulty in enforcing terminations and other franchisors rights. Business failure of the master franchisee can effect the entire franchise system and damage done to its reputation may be difficult to repair. Many major restaurant franchisors choose master franchising as it is cost-effective and the fastest road to international franchising.

With the advancement of technology, educational standards, and economic conditions in many countries, a growing number of management-oriented entrepreneurs has emerged who can serve as master franchisees or be in charge of regional chains. Multimedia teaching tools and other training methods can be effectively used for training purposes.

Joint Ventures

In a joint venture, the franchisor joins with a local investor, thereby forming a joint venture restaurant or chain of restaurants. In this relationship, the franchisor has more control than in the case of a master franchise. Unlike under master franchise agreements, franchisees do not possess the rights to subfranchise or establish units within a particular country or region. There may be several joint ventures in a country, whereas there can be one master franchisee only in a country or region. Franchisors may enter into a joint venture due to the requirement of a particular country or as a means of developing familiarity with the region.

Joint ventures are becoming a popular method for doing business in foreign countries. They require careful assessment and selection of reputable individuals and corporations who possess the necessary experience. Because local political and economic situations can be best understood and handled by local businesses, joint ventures prove mutually beneficial in most, if not all, countries.

The possibility of joint ventures exist particularly in developing, prosperous countries with ambitious entrepreneurs. Franchisors should understand that in some countries there are bureaucratic procedures that require knowledgeable partners with insight and influence in politically active circles. Without inside knowledge, delays in getting minor approvals can take months and sometimes become impossible to secure.

Situations for considering joint ventures were ranked in descending order by restaurant executives as follows:

1. government restrictions on foreign ownership
2. opportunities to make use of the resources of the franchisee in areas of expertise
3. limited exposure to risk while retaining equity in the franchise business
4. widespread use of English language

The advantages of joint ventures in international markets include reducing business risk, increasing production efficiencies, and overcoming entry barriers and better local acceptance. Other advantages include access to resources, flexibility, lower cost of conducting business, and less investment commitment. Joint ventures are sometimes necessary where government control is predominant or a requirement of a country. The advantages also include the fact that the local political environment, economic situations, and regulations can be better understood and handled by local people. Also, the risk of business failure is spread between the partners. The disadvantages stem from the requirement in many countries that joint ventures be structured with more than 50 percent participation by local residents, thereby giving more operational control to the host partner. In case of disagreement or failures, this can pose a major problem. The chances of potential conflict are greater in joint ventures, when compared to other methods of franchising which may be due to different management styles, changing circumstances of conducting business, and control of business.

Decision-Making Model for International Franchising

A generic decision-making model (Figure 10.2), described below, involves six steps. The first step is to plan international franchising. Before proceeding to an analysis of franchising methods, a thorough examination of the environmental factors is necessary as a second step of decision making. In this step, data pertaining to the political, economic, sociocultural, and franchise-related environment of prospective sites are collected. The third step involves the consideration of different methods of franchising. In the next step, the franchising methods are analyzed and evaluated in light of the franchisor's business objectives. Ranking of the benefits of each alternative franchising method and analyses of the environmental data of the selected sites are undertaken in this step. Important factors to be taken into consideration include (a) availability of human resources, (b) training needs of the franchisees, (c) political stability of the country, (d) level of economic activity and availability/accessibility of financial resources to the franchisor, and (e) cultural differences/barriers within the host country. The next step is the selection of the method for franchising that best suits the franchisor's

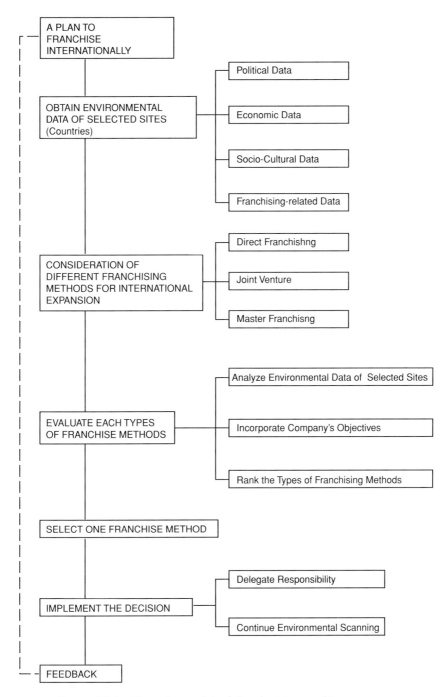

Figure 10.2 Generic model of the decision-making process to select a franchise method for international expansion

needs based on the analysis of the data. Implementation is the last step of the decision-making process. Continuation of environmental scanning and delegation of responsibility are included in the final step. This model can serve as a guide in planning and decision making while selecting a method for international franchising.

Assessment of Environmental Factors

The following checklists are based on our study* conducted to assess factors in selected countries. These factors, which have been modified and presented as a checklist or scoring sheet, help in screening an international environment because they are exhaustive in nature. These factors were selected and ranked by a selected Delphi panel consisting of hospitality industry executives; they appear in descending order. The tables can be used as checklists or for scoring factors.

CHECKLIST/SCORE SHEET FOR ASSESSING POLITICAL FACTORS

Check	Factors to Be Considered	Comments/Score
	registration and enforcement of trade and service marks	
	tax aspects of foreign franchising pertaining to royalties	
	limitation on foreign exchange convertibility	
	government legislation particularly focused on franchising	
	controls on the importation of goods and supplies	
	government instability and uncertainty	

---

* This study was conducted by V. Bosereewong as a part of a doctoral dissertation submitted to Virginia Tech, 1994.

CHECKLIST/SCORE SHEET FOR ASSESSING POLITICAL FACTORS *(continued)*

Check	Factors to Be Considered	Comments/Score
	registration and enforcement of trade and service marks	
	risk of government taking part of business	
	restriction on import tariffs	
	restriction of fair trade and competition with indigenous companies	
	lack of active organized labor unions in the food service industry	
	lack of onerous government social programs	
	ability to own land as a foreign entity	
	prospects for success of franchisee	
	real estate costs and availability	
	ability to find local sources of supply	
	cost and availability of raw materials	
	rapidly growing middle class	
	high level of economic stability and growth	
	business climate affecting franchising	
	a plentiful workforce of young, educated people	
	increasing levels of disposable personal income	
	dollars spent on luxury items versus dollars spent on necessities	
	construction costs	
	wage level	
	increase in number of tourists	
	coping with foreign currency uncertainties	
	sophistication in management systems and business structures	
	high level of taxes	
	inflation and interest rates under control relative to the rest of the world	

CHECKLIST/SCORE SHEET FOR ASSESSING POLITICAL FACTORS *(continued)*

Check	Factors to Be Considered	Comments/Score
	a strong natural resource base that would enable the country to continue to generate wealth	
	rapidly growing GNP that is driven primarily by exports	
	international debts	

CHECKLIST/SCORE SHEET FOR ASSESSING SOCIOCULTURAL FACTORS

Check	Factors to Be Considered	Comments/Score
	dietary restrictions (consumption of beef, pork, etc.)	
	availability of local labor	
	growing, young population	
	a country with a strong work ethics and traditional family values	
	potential for overcoming cultural/language barriers	
	cultural acceptance of a dining establishment as a place to socialize with friends and family	
	high literacy rate	
	familiarity with the English language and American idiom	
	high educational levels	
	increasing number of working women	
	trend toward shortening of the work week and national development of more leisure time	

CHECKLIST/SCORE SHEET FOR ASSESSING /FRANCHISING-RELATED FACTORS

Check	Factors to Be Considered	Comments/Score
	capability of potential franchisees (expertise and financial)	
	acceptance of the restaurant concept/product by local population	

CHECKLIST/SCORE SHEET FOR ASSESSING /FRANCHISING-RELATED FACTORS *(continued)*

Check	Factors to Be Considered	Comments/Score
	ease of royalty repatriation	
	training franchisees	
	ability to support franchisees	
	ownership of proprietary information and technical know-how	
	availability of supplies and specified products that meet the restaurant's standards at reasonable cost	
	availability of qualified franchisees	
	ability to control franchisees through the legal system	
	financial incentive (size of up-front territorial fees payable by franchisees)	
	suitable location for the restaurant concept (such as shopping center, etc.)	
	potential size of the market	
	interface between franchisors and franchisees	
	the success of other franchised American business in the country	
	reports on the operations (such as sales reports)	
	product testing needed to adapt menu items to local tastes	
	an entrepreneurial spirit within the country that encourages people to become franchisees	

All of the above-mentioned factors have to be considered based on the countries selected for business. The method of franchising has to be determined after a careful assessment of the environment in light of the corporation's objectives. Among the empirical data to be collected after a decision has been made to enter a country are:

- the population growth in the past and projected growth beyond the year 2010
- the size of the expatriate population in countries where there are considerable expatriates
- demographic information, particularly about age groups
- growth in the GDP (gross domestic product) for the country
- expected competition between existing indigenous and foreign restaurants
- the number of U.S. franchise restaurants in the region (One such comparison can be based on the population of a country/city and the number of McDonald's or KFC units there—as example.)
- the popularity of type of food in the country, such as hamburgers, chicken, pizza
- the availability of marketing and advertising agencies; promotional methods
- the location of legal firms
- the availability of supplies and equipment
- the presence of utilities and maintenance services
- applicable site selection criteria

The Global Market for Restaurant Franchising

With the changing political scene resulting in a new world order, international business is booming in almost every part of the world. Added to the political situation is the vast population not yet exposed to restaurant franchises or franchising as a method of doing business. Franchising is believed to be a major benefactor of this change, as it is estimated by the International Franchise Association that more than 60 percent of global business will be conducted using franchising. Also, the change in business alignment in various regions of the world is creating new partnership possibilities. For the purpose of this discussion, the world is divided into four corridors that are expected to become major centers for business activities in the future. Global sourcing and other purchasing efficiencies are generating savings that help support McDonald's value initiatives, as customers around the world want value. Cost reduction becomes necessary because conditions are not similar to those in the United States. In the following discussion, examples are given for illustrative purposes only, as an in-depth discussion is not within the scope of this chapter.

Pan-Pacific Region

Economies in the Pan-Pacific Region were the fastest growing in the world until 1997. This region was experiencing the most positive international economic boom. In spite of economic fluctuations, the hospitality industry, particularly restaurants, is growing. In 1990, KFC had 3,000 international units in fifty-four countries, with more than a third in ten of the fifteen Pacific Rim countries. Among the countries of the region, Vietnam, Singapore, Malaysia, Taiwan, Korea, Indonesia, Thailand, Taiwan, and China are changing rapidly as far as economic growth is concerned. Many of these countries have entered into economic collaboration agreements that have become a source of mutual benefit from the international trade point of view. This region is experiencing a growth in population as well as in the GDP of most of the countries.

Among the existing economies, Singapore, Malaysia, Taiwan, Japan, and Korea are on the forefront. It is evident that some countries such as China and Vietnam have opportunities unparalleled in world history. With the development of industry and based on the factors mentioned earlier, all indications point to an economic growth in that region. Vietnam seems to be emerging dramatically and moving toward growth and change. Vietnam's proximity to Thailand, Malaysia, Japan, Korea, Taiwan, and China places it in the heart or hub of economic activity; it has the potential to become a gateway to the more economically mature countries. The first Baskin-Robbins ice cream store and the first McDonald's restaurant opened in 1994–1995. Other U.S. franchisors are taking a careful look at the situation.

Problems include lack of infrastructure, control of government, lack of property ownership rights, lack of historical data, the currency fluctuations, financial uncertainty, and the hesitation of investors as the changes take place. In short, however, the entire region is open for business and restaurant franchising (both local and U.S.) has potential for growth. Success depends on how well plans are made for entering the market and how soon one can get in to overcome the financial situation. Mutual trade agreements among the countries in this region have become daily events. Trade and travel among countries is being facilitated and investment by individuals and corporations within the region is encouraged. The growth of the population and the popularity of American franchise restaurants definitely makes this region attractive for business. People in this region are accommodating and tolerant, which makes it easier to introduce new concepts.

China is reported by McDonald's to be a strong market with a rapidly growing appetite for Western culture and food.

European Common Market

This will be another corridor with growth potential for franchise restaurants. Agreements such as GATT (General Agreement on Tariffs and Trade) will have an impact on the business and human resource potential in this region. The combined business potential of mainland European countries and England will bring a change in this region. Uniform laws pertaining to import and export, relative ease of moving from one country to another, licensing arrangements that are common to all countries, a unified currency, and all other changes will facilitate the development of franchising and more so the growth of franchise restaurants in all of the countries. Restaurant franchises are present in traditionally conservative areas with respect to cuisine, such as Italy and France. Efforts are also being diverted toward franchising restaurant concepts that are indigenous to several European countries.

With the political changes in the former Eastern bloc countries, a new market has opened. Some countries are picking up the pieces and are trying to develop their economies. This may prove a point of time in which restaurant franchises can get a foothold in this region.

Africa and Middle East

With the relative peace in the region, many countries are diverting their attention toward developing their economies. These countries are definitely going to benefit from U.S. restaurant franchises. It should be noted that due to religious and cultural preferences, chicken and pizza franchises have greater potential for success than do beef sandwich franchises. Also, U.S. franchise restaurants are considered a status symbol and many go to these restaurant to socialize with friends and families. There is a steady growth of population in this region and, as seen from the size of the population and the presence of McDonald's and KFC, there is plenty of room for franchise development. Conveniently, most of the population is concentrated in few cities.

From the travel and tourism point of view, this region will develop very rapidly. Jordan is planning to develop beaches and revive tourism in the region. Egypt is full of historical attractions and Morocco has its own charm. These sites will become tourist destinations that in turn will lead to franchising. Another factor that should be taken into consideration is the religious centers in Saudi Arabia and Jerusalem. Millions of religious tourists arrive in various seasons. Fast food and franchise restaurants have a market of enormous growth potential. However, the political stability and attitude toward tourists should be taken into account before making any decision.

Countries that have oil revenues are expanding rapidly; this industrial expansion will result in a need for franchise restaurants. These areas include small kingdoms such as Dubai, Abu Dhabi,

Muscat, Oman, and Qatar. With planned development, considerable investment and many franchises are entering this region. A fact that also is associated with development is the considerable size of the expatriate populations that come from India, Pakistan, Philippines, Bangladesh, and other countries. Their exposure to and familiarity with foods available at franchise restaurants will create a new market. Many of the expatriates are single, without their families with them, and therefore eat most of their meals away from home. This is an untapped market that franchises can utilize as they grow in this region.

Among African countries, South Africa is undergoing major political and economic changes. Many foreign franchises are entering this country, where growth potential is good.

North and South America

As in the case of GATT, the North American Free Trade Agreement (NAFTA) will have an impact on trade and human resources in the United States, Canada, and Mexico. South American countries also include large populations that can be a potential market for franchise restaurants. Mexico and Canada have been at the forefront of franchise development. Of course, their proximity to the United States led to the rapid development of franchises in these countries. The largest number of U.S. franchises is in Canada, but still is potential for further development.

South American countries such as Venezuela, Brazil, Ecuador, Argentina, Guyana, and Chile are all nations with developing economies. This is also the region in which current franchising is limited and restaurant franchises can find a potential market. Based on culture and food habits, menu modifications will be needed to develop both indigenous or U.S. restaurant concepts.

Restaurant Study 15

Domino's Pizza, Inc.

DOMINO'S PIZZA: OVER 38 YEARS OF LEADERSHIP

Domino's Pizza has been dedicated and committed to quality service, product, and delivery excellence for more than 38 years, making it the world leader in pizza delivery. Domino's Pizza has led the industry by dedicating its attention, energy, and resources to one mission: to be the leader in off-premise pizza convenience to consumers around the world.

The company was founded in 1960 by Thomas S. Monaghan and owes its success to a few simple precepts: a limited menu offered only through carryout or delivery. Domino's delivers quality products with a total satisfaction guarantee: Any customer not completely satisfied with their Domino's Pizza experience will be offered a replacement pizza or a refund.

For 32 years, Domino's Pizza maintained a limited menu of Traditional Hand Tossed Pizza and Coca-Cola. In 1992, the company introduced bread sticks, its first nonpizza product in the United States. In 1993, Ultimate Deep Dish Pizza and Crunchy Thin Crust Pizza were introduced nationally. Buffalo Wings are the most recent nonpizza menu addition, rolled out nationally in 1994. In 1996, Domino's Pizza was the first national chain to launch flavored crusts nationwide.

Domino's Pizza, Inc., is recognized as the world leader in pizza delivery, with 1997 sales of $3.2 billion. Domino's Pizza, with 1,700 franchisees, operates more than 6,000 stores throughout the United States and in 64 international markets.

DOMINO'S PIZZA: ESTABLISHING AN INTERNATIONAL PRESENCE

Domino's Pizza International, Inc., a wholly owned subsidiary of Domino's Pizza, Inc., was established in 1982 as a means of establishing international presence for the world leader in pizza delivery. This wholly owned subsidiary, with its 400-member franchise body, currently operates more than 1,670 stores in 64 international markets. In 1997, the international subsidiary contributed $680 million to the company's $3.2 billion in sales.

Master Franchising is the primary method of overseas development for Domino's Pizza. The Master Franchise system allows Domino's to utilize the business expertise of local partners by selling the rights to develop Domino's Pizza in a country or territory. Domino's Pizza stores outside the United States are all franchise-owned.

A Master Franchise Agreement combines the master franchisee's knowledge of the local market with Domino's Pizza's expertise and operating systems. The Master Franchise Agreement eliminates many cultural challenges of overseas development, including language, local government regulations, and trademark registrations.

Domino's Pizza does support adaptation of operating systems and product to accommodate cultural differences including delivery vehicles, store design, and pizza topping selections. Regional topping preferences are the most common cultural adaptation. These include pickled ginger in India, squid in Japan, and green peas in Brazil. The dough, sauce, and cheese are consistent worldwide to ensure the integrity of the product is maintained.

Converting local pizza chains to the Domino's Pizza brand is another method of international expansion for Domino's Pizza. In 1994, an agreement was signed with an 88-unit Australian pizza company to convert its units to the internationally known Domino's brand. This conversion brought Domino's store count in Australia from 25 to more than 100 in two years. Other established regional pizza companies are currently being targeted for conversions in other markets.

Domino's Pizza International is based at the parent company's world headquarters in Ann Arbor, Michigan.

DOMINO'S PIZZA INTERNATIONAL POPULAR REGIONAL TOPPINGS

Pepperoni may be the number-one topping in the United States, but tastes are evidently more eclectic around the world. Domino's Pizza International, with 1,300 stores in 60 markets, has made the following unusual additions to the standard Domino's Pizza menu:

Squid (Japan)
Tuna and Corn (England)
Black Bean Sauce (Guatemala)
Mussels and Clams (Chile)
Barbecued Chicken (the Bahamas)
Eggs (Australia)
Pickled Ginger (India)
Fresh Cream (France)
Guava (Colombia)

(a)

(b)

A typical Dominos' Pizza Restaurant (courtesy Dominos' Pizza, Inc.)

1997 PRODUCT USAGE FACTS

- In 1997, Domino's Pizza sold over 325 million pizzas.
- Pepperoni is the most popular pizza topping. In 1996, over 27 million pounds were used.
- Domino's Pizza used over 174 million pounds of part-skim mozzarella cheese in 1997.

- Over 3 million pounds of pizza sauce were used in 1997.
- 182 million pounds of flour went into Domino's Pizza dough during the year.
- Over 252,000 pounds of olives were used in 1997—27,000 pounds of green olives and 225,000 pounds of black olives.
- Domino's used 6.5 million pounds of pineapple in 1996. Ham and pineapple is a popular combination throughout the world.
- Domino's pizza lovers ate 11.3 million pounds of mushrooms, 5.3 million pounds of green peppers, 8.3 million pounds of ham, 16.8 million pounds of sausage, 2.6 million pounds of bacon, 7.6 million pounds of beef, and 5.7 million pounds of onions.
- 66,000 pounds of anchovies were used on Domino's pizzas last year.

DOMINO'S PIZZA, INC.: IMPORTANT DATES IN HISTORY

1960	Tom Monaghan and brother James purchase "DomiNick's," a pizza store in Ypsilanti, Michigan. Monaghan borrowed $500 to buy the store.
1961	James trades his half of the business to Tom for a Volkswagen Beetle.
1965	Tom Monaghan is sole owner of company, and renames the business "Domino's Pizza, Inc."
1967	The first Domino's Pizza franchise store opens in Ypsilanti, Michigan.
1968	Company headquarters and commissary are destroyed by fire. First Domino's store outside of Michigan opens in Burlington, Vermont.
1975	Amstar Corp., maker of Domino Sugar, institutes a trademark infringement lawsuit against Domino's Pizza.
1978	The 200th Domino's store opens.
1980	Federal court rules Domino's Pizza did not infringe on the Domino Sugar trademark.
1983	Domino's first international store opens in Winnipeg, Canada. The 1,000th Domino's store opens.
1984	Ground is broken for new headquarters—Domino's Farms, in Ann Arbor, Michigan.
1985	Domino's opens 954 units, for a total of 2,841, making Domino's the fastest growing pizza company in the country.
1989	Pan Pizza, the company's first new product, is introduced. Domino's opens its 5,000th store.
1990	Domino's Pizza signs its 1,000th franchise.

1992	Domino's begins national roll-out of bread sticks, the company's first national non-pizza menu item.
1993	Crunchy Thin Crust pizza is rolled out nationwide. The Company drops the 30-minute guarantee in corporate stores and reemphasizes the Total Satisfaction Guarantee: If for any reason you are dissatisfied with your Domino's Pizza dining experience, we will remake your pizza or refund your money.
1994	Buffalo Wings are rolled out in all U.S. stores. The first Domino's opens in eastern Europe in Warsaw, Poland. The first agreement to develop Domino's Pizza in an African country was signed by Specialized Catering Services, Inc.
1995	Domino's Pizza International division opens its 1,000th store. First store opens on the African continent, in Cairo, Egypt.
1996	Domino's launches its web site on the Internet (www.dominos.com). Domino's rolls out flavored crusts, for limited-time-only promotions, nationally for the first time in company history. The company reaches record sales of $2.8 billion system-wide in 1996.
1997	Domino's Pizza opened its 1,500th store outside the United States, opening seven stores in one day on five continents consecutively. Domino's Pizza surpassed $3 billion in sales for the first time in history. Domino's Pizza launched a campaign to update the company logo and store interior with brighter colors and a newer look.
1998	Domino's launches another industry innovation, Heatwave, a hot bag using patented technology that keeps pizza oven-hot to the customer's door. Domino's Pizza opens its 6,000th store in San Francisco, California, in April.

CHAPTER 11

Franchise Concept Development

In spite of its numerous advantages, franchising is not suitable for all types of business. Certain businesses will never be franchisable, while others can function well only as franchises. Many successful restaurant owners fall into the trap of underestimating the difficulties of franchising.

Simply defined, a concept is an idea conceived in mind and concept formation is the process of sorting specific experiences into general rules or classes. In any concept formation there are two processes—first, an identification of important characteristics, and, second, an identification of how the characteristics are logically linked. Thus, theoretically, concepts serve as norms or models that account for the potential of some things to fluctuate in some respects while remaining constant in others. Applying these basic concepts to franchising, an idea or concept becomes a nucleus on which other business aspects are built. This basic idea thus becomes a franchisable model.

In order to be successful, this concept or idea should be based on sound principles that will lead to franchising. Thus, a concept needs testing and retesting before it is suitable for franchising. A restaurant may be the most popular restaurant in town when independently owned but fail when it is franchised. The intricacies of franchising must be understood before getting involved with it. It sounds attractive if one thinks of licensing to other operators and building a business empire.

Because most restaurant chains use business-format franchising, the discussion in this chapter will be focused on concepts for a business-format franchise. Also, the basic features of successful concepts for restaurant franchises will be considered from the point of view of both products and services. Thus, the business concept for franchising can be assessed with respect to menu, layout and physical facilities, service, marketability, and management. Each of these aspects is described separately.

Menu Development

Simplicity

Most successful restaurant franchises have simple menu concepts. Hamburger chains are a good example; the menu consists of a beef patty placed on a bun. McDonald's started in business with this simple concept. It is much easier to work with a simple menu item than with a complicated one because preparation and service are both much easier. Complexity is the primary reason why fancy table-service restaurants cannot be franchised. Complicated menu items and varied methods of preparation are difficult to duplicate. Training also becomes difficult when complex menu items and complicated service procedures are involved.

Imagine that one of the menu items is a cheesecake to be prepared from scratch, the preparation of which has to be taught to fifty other franchisees who will in turn teach it to one hundred other employees. Will consistency of the product be easily maintained? If all the items on the menu are that elaborate, training will become a nightmare and it will be impossible to maintain product consistency.

Replication Capability

The entire concept, from product to service, should be capable of replication on demand without jeopardizing the quality. A sandwich bought in California from a franchise restaurant should look and taste exactly the same as one bought from the same restaurant in New York or elsewhere. The entire concept of franchising thrives on consistency in both quality and service. The restaurant design, the menu offered, and the service should be consistent among all franchises within a system. No matter how sophisticated and popular an item is, if it cannot be consistently replicated, it is not suitable for franchising. Most ethnic menus and other elaborately prepared menu items cannot be used successfully in franchised restaurants, no matter how popular they are locally. These types of items are more suitable for independent restaurants.

Ready Availability

If a menu concept is simple and can be duplicated easily, the next criterion is that it should be readily available at short notice whenever needed. In franchise restaurants, a busload of consumers may walk in and have to be served within a short period of time, with consistent quality. Thus, the product should be available in the shortest possible time needed for service. Modern technological advances, such as microwave and infrared heating methods, are making this possible. Preparation procedures should be simple and easily understood by any employee.

Quality

The quality of products and services should be stable all year round. Products should be such that they maintain quality and consistency after preparation. Procedures, seasonal variations, and location should not adversely affect the quality of the finished product. Many menu items do not remain consistent but change under varying conditions of temperature and humidity. Nutritional, sensory, and sanitation quality of the product should remain consistently high under all conditions of production and service.

Availability of Ingredients

All ingredients required for the preparation of menu items should be readily available. A product may meet all other criteria, but if ingredients are not readily available, that may hinder its use. Not only must products be available, they must also meet certain specifications. For example, potatoes for french fries should be of a standard variety and have specific characteristics. Product performance is largely dependent on using precise ingredients in exact quantities. Availability in large quantities needed for production is essential. Seasonal variations should not have any adverse impact on the availability of ingredients.

Food Characteristics

An acceptable product should have taste and appeal and should be preferred by large sections of the population, thereby assuring a reasonable market. Products that are highly liked by some and disliked by many, such as lamb and liver, are not likely to be successful. Food characteristics, including their organoleptic (sensory) properties, play an important role in their acceptance. The most important food characteristics are discussed below.

COLOR   Interesting and coordinating color combinations help in the acceptance of food and, to an extent, help stimulate appetite.

Color combinations also help in the marketing and advertising of food items in print and on television. Careful planning and coordination are essential so that the consumer finds food appealing when taken out of a bag or when placed on a tray, counter, plate, or salad bar. Garnishes can make an enormous difference in food appearance. Colorful foods have an eye appeal, which helps make food popular. Color also helps emphasize variety and the presence of different food groups within a presentation. Bright and varied combinations add to foods attractiveness, while dull or monotonous colors convey a sense of blandness. Remember that a consumer often selects food first by eye appeal, so color coordination is an important consideration.

Colors also have a psychological impact on consumers appetite. Red, orange, peach, pink, brown, yellow, and light green are considered desirable colors for foods. Purple, violet, dark green, gray, and olive are less desirable colors. Although artificial coloring may be added to enhance food colors, natural food colors are preferred. Fruits and vegetables, in all variety of forms and shapes, add color to the menu.

TEXTURE AND SHAPE   The textures and shapes of foods affect consumers preferences. Certain foods are preferred because of their hard texture and some because of their soft texture. A desirable combination of soft- and hard-textured items on a menu is essential. Impressions of the texture and shape of a particular food are formed even before tasting it. Texture can best be detected by mouth feel. *Soft, hard, crispy, crunchy, chewy, smooth, brittle,* and *grainy* are some of the adjectives used to describe food texture. Certain combinations of food items go well together.

CONSISTENCY   *Consistency* refers to the degree of viscosity or density of a product. Like texture, consistency provides for variety among menu items. *Runny, gelatinous, pasty, thin, thick, sticky,* and *gummy* are the most common adjectives used to describe consistency. Food items should have a desirable consistency. The moisture content of food directly affects its consistency. Foods that include ingredients of low consistency may be difficult to package, handle, and prepare for take-out. The consumer may find it inconvenient to handle them for fear of messing things up. When mayonnaise, gravies, sauces, and ketchup are used, consistency should be taken into consideration.

FLAVOR   For obvious reasons, the flavor of food should be given prime consideration in the selection of menu items. Foods can have sweet, sour, bitter or salty flavors, present alone or in combination. A desirable blend of flavors is essential for creating a successful innovative food. The predominance of any one flavor is usually undesirable. A contrast in the types of flavors and their intensity adds to the acceptability of menu items. Bland foods may be made more appetizing by adding pungent sauces, or a blend of sweet and sour flavors may be tried. Barbecue flavors are also desirable in certain types of foods. Pickles and mustard add to the flavor of foods when used in proper quantities.

Adequate testing should be done so that these supplementary flavors enhance the flavor of the menu item rather than mask it. The right combinations of spices and condiments are essential in order to develop the right kinds of flavors. Frequent taste testing and standardization are needed to achieve desired flavor combinations. Artificial flavor enhancers should be avoided as much as possible due to the allergic reactions that some are linked with. In vegetable cookery, strong-flavored vegetables should be combined with mild-flavored ones. For example, onions, broccoli, cabbage, and peppers have strong flavors and therefore should be complemented with milder vegetables whenever combinations are used. It is advisable, again, that these vegetables enhance rather than overshadow the flavor of the primary food item. From the nutritional point of view, vegetables provide valuable fiber, which should be taken into consideration. Strong-flavored sauces and gravies go well with bland items, such as mashed or baked potatoes and spaghetti.

Last, but not least, when considering flavor is the odor of the food. Food items should have a mild and desirable odor, preferably signifying the method of preparation. For example, a barbecue or smoked odor is desirable for some foods. Odor may serve as a marketing tool in itself. The scent of fresh-baked cinnamon cookies in a shopping center provides instant appeal.

Preparation Method Consideration should be given to the method of preparation used. There are many ways to prepare foods. Careful consideration is necessary because the method selected determines the kind of equipment needed. It is advisable to select a method that can be used for a variety of menu items. Methods of preparation include frying, baking, broiling, boiling, steaming, grilling, braising, or a combination of these. In each of these types of preparation, several variations are possible. Offering fried, grilled, and roasted foods, for example, provides choices for the customer. Choice becomes important as competition among franchise restaurants becomes more intense. It also helps in building repeat business. From the management point of view, it is essential that there be uniformity in methods of preparation in order to facilitate optimal use of employee skills and equipment.

Methods selected should have processes that are not complex and that do not take a long time. Simple and straightforward methods are easier to understand and make training employees easier.

Many franchise restaurants feature a method of preparation as their specialty and even demonstrate the process while food is being prepared. Hamburgers or steak flame-broiled in front of a customer have a distinct appeal. Many baked goods and confectionery items are prepared so that the customer may view the entire preparation process. The show adds to the marketability of the products.

Serving Temperature Serving temperatures should be well controlled. An item may be consumed within the restaurant, carried away from a drive-through for later consumption, or delivered.

Foods selected should be such that they retain the desired temperature as much as possible at the time of consumption. Greasy items that get cold may not be desirable. There is no strong evidence that season or weather affects the selection of particular temperatures of food, but it may be desirable to include both hot and cold items. For example, milkshakes, ice cream, and salads may be served to complement hot entree items. Also, the time and method of consumption should be considered.

PRESENTATION   The final appearance of food—whether on a plate, cafeteria counter, serving tray, food container, bag, display case, or take-out package—is important. Neatly presented foods have their own appeal. Presentation should be carefully planned so that the quality of each menu item is optimal as it reaches the hands of the consumer. The appearance of home-delivered menu items should not be overlooked. Another aspect that should be considered is the packaging of foods, whether for consumption at the restaurant or for the delivery at home.

Nutritional Quality

The nutritional quality of food is becoming more and more important to consumers. If, in the initial stages of concept development, plans are made for adequate nutritional quality, a lot of criticism and concern will be avoided at a later stage. Diets should be based on a variety of common foods in order to provide all nutrients. Nutritional labels are printed on food packages and give information pertaining to calories per average serving; the amount of carbohydrates, proteins, and fats; and the percentage of nutrients provided in significant quantities by the specified normal serving of that food. An example of a nutrient label is shown in Figure 11.1.

Nutritional labels for final products are desirable because they illustrate to interested consumers the amounts of nutrients and calories in an average serving of the items. Nutritional analysis should be conducted during the testing stages and the final label should be used. If facilities for nutritional analyses are not available, outside consultants should be employed. Growing nutritional concerns among consumers have caused many food service operators to assess menu items and portion sizes. Salt, sugar, fiber, and

FIGURE 11.1 An example of a nutrition fact label

overall carbohydrate content provided by items needs careful assessment when planning menu concepts. For specific nutrients, such as sodium, there are guidelines for recommended ranges of safe intake. A variety of foods, properly used, will provide most of the necessary nutrients. Related to the nutritional content of foods in menu planning are consumer food preferences, discussed later in this chapter.

Another way of looking at the nutritional content of food is the 1996 Dietary Guideline for Americans. Based on these guidelines, the menu concept can be tested and some of the ingredients required may be selected. Some measures that meet the guidelines are described below as examples:

- *Eat a variety of foods.* Whether a sandwich, salad, or any other entree, food selection may be made in such a way that a variety is provided. Meats, vegetables and fruits, grains and cereals, and dairy products should be carefully selected to complement each other. This may involve the selection of lean meats, fish, and poultry, legumes, green and colored vegetables, enriched breads, and dairy products, such as cheese and yogurt.
- *Maintain healthy weight.* Smaller portions and a selection of healthy items help in meeting good nutritional guidelines. Less fat and fatty products, less sugar and sweet substances, and limiting portion sizes may be used to help consumers achieve this goal.
- *Choose a diet low in fat, saturated fat, and cholesterol.* This guideline can be achieved by
    a. selecting lean meats, fish, poultry
    b. selecting plant sources of food whenever possible, particularly protein-rich legumes
    c. moderating use of eggs, organ meats, and seafood, which are high in cholesterol content
    d. limiting the use of butter, cream, animal shortening, lard, coconut oil, tropical oils, and other related products

e. trimming fat off meats whenever possible
　　f. using cooking processes such as broiling, baking, and steaming whenever possible and avoiding deep-fat frying
　　g. reading food labels of ingredients and products used in food preparation
- *Choose a diet with plenty of vegetables, fruits, and grain products.* As discussed earlier, it is important to provide a wide selection of foods. Fiber, which is highlighted in these guidelines, has several health advantages. It can be provided by selecting fruits and vegetables, whole wheat bread, and oatmeal.
- *Use sugars in moderation.* Cutting sugars helps to reduce calories. This can be done by a careful review of products containing sugar. Measures like using natural substitutes for sugars, cutting portion size, and, in general, reducing the level of sugars in food products are desirable. Desserts in the form of light items containing fruit or yogurt may be substituted.
- *Use salt and sodium in moderation.* Sodium intake by Americans is considerably high, and there is a need for curtailing sodium intake. A misconception is that salt is the only source of sodium in the diet; there are several other sources in addition to the presence of sodium as a component in processed foods, snacks, beverages, seasonings, and soft drinks. Salt can be replaced by other flavors, such as lime and lemon juice. Reducing the level of salt in food items is highly desirable. The sodium content of spices, condiments, flavor enhancers, meat tenderizers, etc., should be checked. Soy sauce and MSG (monosodium glutamate) contain sodium and should be used with caution.
- *If you drink alcohol, do so in moderation.* Alcohol adds to calories and is low in nutrient content. Alcoholic beverages and the items that go with them should be carefully screened before selection.

All the abovementioned guidelines should be considered in the development of a healthy menu.

Food Preference

Any menu concept that is successful is based on the food preferences and acceptance of its consumers. Food likes and dislikes are acquired habits. The target population should have a preference for or an acceptance of the menu items. A food preference form may be used prior to or during the testing of a menu item. An example of a food preference survey questionnaire is shown in Figure 11.2. Food preferences should be carefully weighed before selecting a product.

Packaging and Delivery

The size, shape, contents, and consistency of food should be such that it can be packaged and delivered when necessary, whether to the home or via the drive-through window. Materials used for packaging should not have an adverse effect on the quality and safety of foods. Packaging material for delivery should be such that the desired temperature of foods can be maintained as long as possible. Also, these materials should be easily degraded or recycled, keeping in view the preservation of the environment.

Cost-Effectiveness

Finally, the cost-effectiveness of menu items should be evaluated. A product may have all other attributes but may not be cost-effective when food and labor costs are taken into account. Franchising looks for volume production in limited time. Thus, the costs involved need careful evaluation. Items like steaks and shrimp are definitely popular, but their cost-prohibitiveness for an operation should be assessed carefully before final selection. Further, there should not be the possiblity of wide price fluctua-

Name: _____

Using the following scale, circle the number that most appropriately describes your like or dislike for the entrées listed below. Please try to answer as accurately as possible. Thank you.

0	1	2	3	4	5	6	7	8	9
Never Tried	Dislike Extremely	Dislike Very Much	Dislike Moderately	Dislike Slightly	Neither Like Nor Dislike	Like Slightly	Like Moderately	Like Very Much	Like Extremely

ENTRÉE	HOW MUCH YOU LIKE OR DISLIKE	COMMENTS
1. Carrot, Cheese, and Rice Casserole	0 1 2 3 4 5 6 7 8 9	
2. Cheese, Bacon, Tomato on English Muffin	0 1 2 3 4 5 6 7 8 9	
3. Cheese Fondue	0 1 2 3 4 5 6 7 8 9	
4. Cheese Rarebit on French Bread	0 1 2 3 4 5 6 7 8 9	
5. Baked Cheese Sandwich	0 1 2 3 4 5 6 7 8 9	
6. Waffle Cheese Sandwich	0 1 2 3 4 5 6 7 8 9	
7. Cheese Soufflé	0 1 2 3 4 5 6 7 8 9	
8. Egg Croquette	0 1 2 3 4 5 6 7 8 9	
9. Barbecued Hamburger Squares	0 1 2 3 4 5 6 7 8 9	
10. Quiche Lorraine	0 1 2 3 4 5 6 7 8 9	
11. Sausage Patties w/Gravy	0 1 2 3 4 5 6 7 8 9	
12. Broccoli, Tomato, and Jack Cheese	0 1 2 3 4 5 6 7 8 9	
13. Chicken Livers w/Rice	0 1 2 3 4 5 6 7 8 9	
14. Roast Beef	0 1 2 3 4 5 6 7 8 9	

FIGURE 11.2 An example of a food-preference survey questionnaire

tions for the menu items or for the ingredients used in their preparation. The bottom line is that the item should be profitable and have a built-in profit margin.

RECIPE TESTING AND STANDARDIZATION

Recipes that are tested for quality, quantity, procedures, time, temperature, equipment, and yield are called *standardized recipes*. Recipes are standardized to the extent that when the specified ingredients, conditions, and procedures are adhered to, the result is always a predictable product of uniform quality. Standardized recipes assure quality control and are effective management tools. Standardization calls for careful assessment, testing, and evaluation of recipes before their final adoption.

Standardized recipes offer the following advantages:

1. Provide detailed information pertaining to ingredients, procedures, and equipment to be used; thus, a change in personnel does not affect the food quality or quantity.
2. Facilitate cost analyses of the recipes.
3. Provide predictable food quality.
4. Facilitate food purchasing, as exact quantities of food and ingredients to be used are known.
5. Permit easier portion control, if serving portions are listed.
6. Help to set selling prices for menu items and facilitate price changes when the costs of ingredients change.
7. Help in work scheduling, as time and procedures used are standardized. Efficient scheduling results in an even distribution of work and job satisfaction.
8. Contribute to consumer satisfaction, as quality of food and serving sizes are uniform.
9. Avoid confusion and reduce the chances of poor handling and food preparation failures, which could prove expensive for the operation.

10. Help train employees in good production and handling procedures.

Standardization Procedures

The standardization of recipes demands careful evaluation and testing. Standardization involves the adjustment and readjustment of ingredients and their proportions to produce the most acceptable quality. This calls for subjective as well as objective evaluation. Taste tests should be conducted several times until high-quality products are assured. Most often the recipes are enlarged and tested from smaller quantity recipes. The following steps represent the recipe enlarging and standardization procedures:

Step 1. The first step is to prepare the recipe in the minimum quantity for which it was intended. The finished product should then be evaluated. The smaller recipe should be evaluated first on the basis of preparation method; ingredient proportion; availability of ingredients; cost, yield, equipment, skill, and abilities of personnel; and its overall suitability for the operation. A careful screening at this point will eliminate expensive large-scale testing at a later stage.

Step 2. The original recipe should be multiplied and specifically tested. In enlarging, ingredient ratios play an important role. For example, the ratios between sugar and flour, and flour and shortening, are very important. The physical form of the ingredients should be carefully assessed. Fresh chopped onions may be responsible for flavor in a smaller recipe, but it may not be feasible to use chopped onions in larger quantities, as this may change the taste of the finished product. Salt and seasonings need very careful assessment because simple multiplication never works for them. Recipes can be enlarged and standardized if the operation calls for large quantities.

Consumer Surveys and Taste Evaluations

Recipe evaluation should be considered the most important task, because the profitability of the restaurant depends on it. Several franchisors maintain test kitchens in their research and development departments, where recipes are tested and evaluated on a regular and continual basis. Because several factors have an impact when food is delivered from the kitchen to consumers, these assessments provide valuable feedback. A taste panel of experienced personnel within the franchisor's corporate office may be selected to evaluate prepared products. Because there may be changes in food items delivered (brand names and seasonal variations), it is necessary to conduct several replications and periodic evaluations. A form for such evaluations is given in Figure 11.3. It is also advisable to have consumer surveys and taste evaluations whenever new items are introduced—and even for items that are

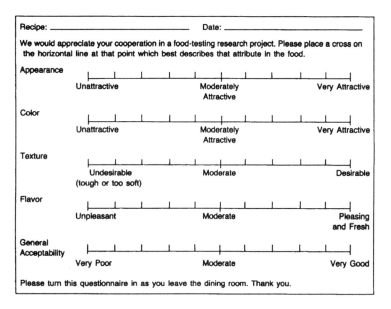

FIGURE 11.3   An example of a taste panel scoring sheet

already being served. Samples may be given out or sold in test market franchise units. Testing may generate interest among consumers as well as provide a marketing opportunity. A form for such an evaluation is given in Figure 11.4. Consumer evaluation should be as simple as possible and contain a limited number of questions. Only those questions that are pertinent and can provide usable answers should be included. Results should be carefully tallied, statistically analyzed, and studied. Action should be taken and decisions made after a careful review of these surveys.

We would appreciate your cooperation in a class food-testing project.

Name of product: _____

1. How would you rate this product?
   Excellent ☐ Very good ☐ Good ☐ Fair ☐ Poor

2. Have you tasted this or a similar product before?
   Yes _____ No _____ (If yes, where? _____)

3. Was your selection affected by:
   (a) Serving person _____
   (b) Person in line before you _____
   (c) Person who invited you _____
   (d) Recommendation from others _____
   (e) Other reasons (please mention) _____

4. Please rank the reasons for selecting this product using numerals.
   (a) Name of the product _____
   (b) Appearance _____
   (c) Previous experience _____
   (d) Price _____
   (e) Serving size _____
   (f) Curiosity _____
   (g) Taste _____
   (h) Nutritive value _____
   (i) Vegetarian nature _____
   (e) Other (please mention) _____

5. Will you try this product again? Yes _____ No _____

6. Other comments: _____

Please turn this questionnaire in at the cashier's desk. Thank you.

FIGURE 11.4 An example of a consumer acceptance survey form

Layout and Physical Facilities

A major aspect of the entire business-format franchise is the layout and physical facilities of the restaurant. The general functions of a restaurant operation are receiving, storage, preparation, service, and sanitation. These functions may be broken down into subfunctions. The fundamental criterion in planning a restaurant operation is to understand and visualize each and every function within that facility. Each function is carried out by performing several tasks, and each one of these tasks should be considered. The next step is to arrange the facilities for these tasks in a way that allows a smooth and sequential flow within each area. It is imperative to develop a flow diagram in order to plan the physical facilities and space allocation for each function. Some of the common functions for which space allocation should be considered are described below.

Receiving Area

The receiving area for foods, beverages, and supplies must be planned for maximum efficiency. The type and frequency of deliveries play an important role in planning. The number of deliveries, in turn, is determined by sales, availability of personnel, space in the storage area(s), and other related operational aspects. Adequate space is needed for receiving, checking, moving, stacking, and transporting items received. Space for much-needed equipment, such as scales and inspection tables, should be planned.

Loading docks should be planned on the basis of the type of deliveries expected. The truck-bed height should allow goods to be transferred efficiently from trucks to storage areas by carts or other mobile equipment. The delivery platform should be of convenient height and preferably about 8 feet deep. Movable platforms are preferable so that the heights can be adjusted. Gravity slides, conveyors, electric cars, elevators, and escalators may also be used for transport from the receiving areas. The receiving area should be located away from and preferably out of sight of the main guest entrance or dining area. It should be easy to find, but preferably away from congested traffic areas. Facilities for disposal of cartons, boxes, and other wastes are necessary. Some provision for limited temporary storage is desirable.

Storage Areas

Storage areas vary according to the types of foods and beverages being stored and the temperature required for their storage. Sanitation and safety should be considered in planning all storage areas. Because quick and easy location of all stored products is highly desirable, they should be arranged conveniently and in a logical sequence. Standard patterns are recommended for uniform arrangement within storage areas. Stacks should be neatly arranged and easily movable. A well-organized storage area reduces movement as well as labor and material costs.

The space required for storage areas is dependent on:

- Type of menu
- Temperature and humidity requirements
- Frequency of deliveries
- Maximum volume required to be stored
- Desired duration of storage

Ample aisle space is needed for the efficient movement of foods and personnel.

DRY STORAGE AREAS Dry storage areas are used for foods and various supplies to be kept at a temperature range of 50° to 70°F (10° to 21°C). The desired relative humidity for this area is approximately 50 percent. This area should preferably not be located in the basement or near heat-generating equipment, such as motors or compressors, or near steam pipes. These areas are not suitable for dry storage because their temperatures may be undesirable for the goods stored. Also, any leakage or sweating may lead to spoilage of stored items. Adequate ventilation is needed for such foods as root vegetables and unripe fruits. Certain supplies, such as linens, towels, paper goods, glassware, silverware, and furniture may also be stored in the dry storage area. However, detergents and other cleaning supplies should be stored away from food items.

REFRIGERATED STORAGE AREAS Refrigerated areas are needed for products that have to be stored at temperatures of 35° to 40°F (1.7° to 4.4°C). These temperatures are essential for storing fresh meats, fruits, vegetables, dairy products, leftover items, and beverages. Refrigeration space is also used for thawing meats. Walk-in and reach-in refrigerators are ideal for refrigerated storage. Walk-in refrigerators are desirable for foodservice operations serving three hundred to four hundred meals per day.

Generally, a refrigeration space requirement of 15 to 20 cubic feet per one hundred meals is recommended. Storage shelves should range from 2 to 3 feet in width. Aisle widths should be a minimum of 3 feet. Space requirements for refrigerator storage areas depend on the volume and type of products to be stored and the number of meals served per day.

FROZEN STORAGE AREAS Frozen storage areas are required for food storage at temperatures of −10° to −20°F (−23° to −28.9°C), and are mainly suitable for storing frozen food items. Predictably, there will be a considerable increase in the use of frozen food items, so the frozen storage facility should be very carefully planned. As in the case of refrigerators, freezers may be of the walk-in or reach-in type. A walk-in freezer may be desirable for a food service operation serving three hundred to four hundred meals per day. The space requirements for frozen storage may be less than for refrigerated storage. Normally, both walk-in freezers and walk-in refrigerators are made as one unit, with an entrance door leading to the walk-in refrigerator and a separate door to the freezer unit inside the refrigerator. This helps conserve energy, as each time the freezer door is opened, the cold air from it goes into the refrigerated area, which is desirable. Sufficient aisle space and space for the door as it opens should be provided. In order to conserve energy, it may be advisable to have plastic curtains or strips at the entrance.

Preparation Area

Many restaurant operations have areas for the pre-preparation of certain items, such as peeling and coring of vegetables and thawing and trimming of meats. The space allotted to this area is dependent on the number of meals served and the amount of pre-preparation needed at the facility. In some operations, this space is also used for ingredient weighing as well as such preliminary preparations as mixing and marinating. Sinks, water lines, and other equipment need to be included in this area. The preparation area or production area is the principal activity center of any restaurant operation and normally has several functions. Usually, areas are assigned for the preparations of different foods. The amount of space required for each of these functions is dependent on the menu and the number of meals served per day. Based on the space available, these functions may be combined. Many restaurants have to utilize available and often limited facilities for carrying on all functions. Space should therefore be allocated based on the importance and priority of the function.

Space required for preparation areas is based on (1) the type of menu and the items included in the menu, (2) the type of preparation and the extent of the required preparation, (3) the number of items and the quantity of each item to be prepared, (4) the type of service, and (5) the equipment available for preparation.

It is evident that the space requirements for various types of restaurant facilities are based on numerous factors. Once the areas for different types of preparation are designated, work spaces should be designed, incorporating principles of work engineering and such human aspects as height and reachability.

A work area 4 to 6 feet long is usually convenient for an average person. The width of the worktables may range from 24 to 30 inches A height of 34 to 36 inches is adequate for worktables. However, whenever possible, adjustable table heights are desirable. All these factors should be taken into consideration while planning, as they directly affect the productivity of employees and reduce their fatigue.

Aisle spaces, which may range from 36 to 42 inches in width, should be planned between work spaces and be wide enough to permit the movement of carts. They should allow free movement of traffic without interfering with work. Aisle should, preferably, be perpendicular to work areas and limited to a minimum, as from a service point of view these are nonproductive spaces. Aisles should be used only for the traffic associated with service or production linked to food preparation. Through traffic should never be allowed in work areas.

As a general rule, equipment in the preparation area should occupy a maximum of 30 percent of the total area. It is important to plan the placement of equipment in the preparation area as well as to plan the work spaces. All equipment should be placed in such a way that there is easy access without much movement by the personnel working in the kitchen. Normally the production area should be attached and conveniently located to the serving area.

Templates, which are small models or scale drawings, are useful in planning the layout and design of the facilities.

Serving and Dining Areas

Serving and dining areas are dependent on the style of service employed by the particular restaurant operation. Because there are various styles of service, there is a variety of plans to match. Most service areas are designed to take into consideration the likes and dislikes of the food patrons, with particular emphasis on atmosphere.

In most restaurants, the service areas are attached to the preparation areas and there is a pick-up center through which food is served to the customers. Particular emphasis should be given to planning so that the dining area is not affected by the smoke, heat, and sounds of the food production areas. Preferably the door of the preparation area should not open directly into the dining area. Space should be provided for hot holding of items such as french fries and apple pies or display of cold items such as salads.

Estimating the space requirements for dining areas may be difficult, as different types of restaurant operations require different space allowances. In general, the space requirements for dining areas are based on:

- Type of service
- Number of consumers served per meal
- Largest number of consumers served at one time
- Type of menu offered and number of choices
- Turnover rate (Turnover rate per hour is the number of times a seat is occupied in one hour. Turnover rate per hour multiplied by the number of seats available gives the number of patrons served per hour.)
- Type and pattern of seating arrangement
- Table and seat sizes, shapes, and numbers
- Aisle spaces between seats
- Number and location of service stations
- Special space required for beverages, napkins, condiments, straws, trash containers, etc.

Serving and dining facilities should be planned using all the above-mentioned factors. The best possible plan for a dining room takes into consideration the maximum number of consumers that can be accommodated within the dining area without disruption of service or inconvenience to patrons. Safety of everyone within that facility should be considered. Sufficient exit areas as well as aisle spaces for movement are essential. There should be adequate spacing between rows of chairs. Service aisles should be about 3 feet wide and the access aisle (for the consumers) should be 1.5 to 2.0

feet wide. At least one aisle for patrons should be wide enough to allow for the passage of wheelchairs.

In general, the comfort of consumers should be given primary importance. For comfort, adults require approximately 12 square feet of space. However, this may vary based on the type of foodservice operation. At some restaurants, fast turnover is needed, and at such places seating that is too comfortable may not be desirable. Airports, where people may have to wait for a considerable period of time, are a good example of places where people may use eating establishments for sitting and spending time. For that reason, restaurants may have stand-up food bars or less comfortable seats to discourage their use as waiting areas. The same is true for certain quick-service operations, where seating is planned to encourage rapid turnover. Sufficient elbow room should be provided for the comfort of consumers as well as for smooth service by food servers.

Seating in a dining room should be planned so that a maximum number of customers can be served without inconvenience. Service stations should be planned at convenient locations so that the maximum number of consumers is served in the minimum possible time. These areas should have space for storing foods and equipment prior to and during service. Space must be provided for temporary storage of soiled dishes or trays.

Sanitation Area

Sanitation areas include the dishwashing and pot-and-pan-washing facilities. Space requirements vary based on (1) the volume of dishes, pots, pans, and other utensils to be cleaned at one time, (2) space available for holding clean dishes and utensils, (3) type of washing facilities and dishwasher, (4) methods used for cleaning and sanitizing, and (5) the number of personnel available for work.

The sanitation area should be organized so that there is room to stack clean dishes and utensils. In addition to dishwashing and pot-and-pan cleaning, the area should include garbage disposal space and a place for keeping mops, brooms, and other housekeeping equipment. The space allocated for sanitation is primarily dependent on the type of restaurant. Public health department regulations must be considered when planning this area.

In addition to all of the abovementioned areas, space should be allotted for employee facilities, such as lockers, lounges, toilets, showers, rest rooms, and office space. The space allowance for these facilities is based on several factors, including the type of restaurant operation.

Equipment Selection and Layout

Selection of the proper equipment is extremely important. The type of equipment selected depends on the area in which it is to be used. Because equipment represents a fixed asset of a foodservice operation and depreciates the moment it is purchased and installed, equipment selection requires careful planning and decision making. If improperly selected, equipment may tie up much-needed cash and lead to the failure of the operation. A variety of restaurant equipment is available with varying degrees of modifications and a wide price range. Whether a particular piece of equipment is really needed and whether it is a good investment is the most difficult decision a franchisor must face. Careful calculations are needed before such a decision is made. The volume of food production and handling, employee productivity, and the profitability of the restaurant directly depend on the type of equipment available at that facility. The factors to be considered in the selection of equipment are discussed below.

NEED It is obvious that need should dictate the purchase of any equipment. Need and planned use of equipment should be evaluated on the basis of whether or not the purchase or addition of that particular equipment will result in (1) desired or improved quality of food, (2) significant savings in labor and materials costs, (3) increased quantity of finished food products, and (4) if it will contribute to the overall profitability of a restaurant operation.

Essential equipment should be given priority and preliminary selection should be based on the basic needs of an operation. Need for a particular piece of equipment should be assessed from several points of view as, for example: (1) whether the equipment will be used for prolonged periods of time, (2) whether it has the potential for meeting the future needs of the foodservice operation, (3) whether the equipment will require maintenance, (4) whether there exist alternate, less expensive versions of similar equipment that can adequately meet the demand of the operation, and (5) whether the equipment can be modified for multipurpose use. In other words, the need should be calculated and well established. It is not financially advisable to buy expensive equipment or larger, more sophisticated equipment than that which is essential for an operation.

COST Several costs are incurred in the purchase, installation, and maintenance of equipment in any type of restaurant operation. The major costs incurred in the purchase of equipment are (1) purchase or initial cost, (2) installation cost, (3) insurance costs, (4) repair and maintenance costs, (5) depreciation costs, (6) initial financing costs, interest, and other charges, (7) operating costs, and (8) costs of benefits and losses derived by addition of the equipment. The installation of equipment may require extensive remodeling that may be more expensive than the cost of the equipment itself.

Market prices of equipment vary based on the type of the equipment, its manufacturer, and its utility. A comparative assessment of these costs is essential before making any decision regarding purchasing of equipment. Some of these costs are calculated by manufacturers and are available for consideration before purchasing or making a decision to purchase. Expensive equipment, like dishwashers, requires more careful assessment than relatively less expensive equipment. There are various methods of calculating the profitability of equipment based on costs.

Because special equipment may be needed for a franchise, it is advisable to have that custom made by contacting an equipment manufacturer. It is in the interest of the manufacturer to design equipment and its layout, as it can be used in several franchise units.

FUNCTIONAL ATTRIBUTES   Because equipment is selected to fulfill specific functions of a restaurant operation, it is necessary to evaluate each piece of equipment based on its functional attributes in the light of the desired needs. Performance of the equipment should be assessed based on its cost as well as the availability of other equipment. The broadest possible functionality should be weighted more heavily than costs. The possibility of modifying the equipment by attachments and other changes should be considered an asset. Anticipated menu changes also dictate the type of equipment to be selected. Other considerations include the type of energy required for the operation of the equipment as well as how heavily it will be used. Several types of construction materials are used; all of the attributes mentioned in this section should be considered while planning construction.

SANITATION AND SAFETY   Sanitation and safety are primary considerations in purchasing any equipment for a foodservice operation. Ease of cleaning and sanitation should be given high priority when selecting equipment. No matter how sophisticated a piece of equipment is, if it cannot be cleaned properly, it is not suitable for a foodservice operation. All materials used in the manufacture of the equipment, particularly food contact surfaces, should be made of nontoxic materials. Equipment used in food production should be such that safe temperatures can be maintained easily and at all times. Accessories such as lights and dials should be integral parts to ensure that the proper temperatures are reached and maintained by the equipment. All parts should be easily accessible for cleaning. Wherever applicable and possible, equipment should be able to be disassembled and reassembled easily and quickly to facilitate cleaning. Self-cleaning types of equipment are preferable. The National Sanitation Foundation (NSF) certifies foodservice equipment that meets defined sanitary standards. Whenever possible, equipment that has been approved by NSF should be given preference.

All equipment selected for a foodservice operation should have built-in safety features. Electrical equipment should be used with proper voltage and free from hazards. All moving and sharp parts should be properly protected. There should be no rough or sharp edges on any equipment. Safety locks and devices should be used on all equipment wherever possible, such as in steamers, steam kettles, and carts. Equipment should be free of crevices and holes that may either harbor insects and microorganisms or hinder proper cleaning and sanitation. All factors related to sanitation and safety should be checked before making any decision regarding purchase.

SIZE, APPEARANCE, AND DESIGN   The size of the equipment should be such that it can easily be accommodated in the space available in the layout and design of the facility. It may be difficult to place equipment where limited space is available. Improperly located equipment is a continuous source of inconvenience and even a hazard. The doors and openings of the equipment should be designed in such a way that they do not cause any problems or hazards.

The design and appearance of the equipment should be attractive and should allow maximum performance with a minimum of problems. Durability should be given consideration in the design, as the equipment used in foodservice operations is subjected to constant use and abuse. Although functional characteristics should be given priority over design, the latter should be such that it facilitates smooth functioning. The appearance of the equipment should be congruent, with the type of facility, with matching colors whenever possible. Size, appearance, and design becomes more important if equipment is visible to the customer—the more so if it is intentionally designed as an exhibit

OVERALL PERFORMANCE   Restaurant equipment should be selected on the basis of its overall performance, including such aspects as quiet operation, easy mobility, remote control operations, computerized controls, variety of functions, availability of parts, and ease of maintenance. Those pieces of equipment known to give superior performance should be given priority. It may be more advisable to purchase equipment with proven efficiency rather than relatively new equipment with no history of use. The energy conservation features of the equipment should be considered.

Planning the Overall Atmosphere

Once layout and design are finalized, the next step is to plan the atmosphere, taking into consideration both the future food patrons and the employees. Atmosphere provides the overall impression of a restaurant. Planning involves the creative efforts of architects, engineers, interior designers, and foodservice managers. The atmosphere of a franchised restaurant is practically a trademark in itself and helps build the image. Developing repeat consumers is dependent on how well this atmosphere is received. Many factors must be considered when planning the atmosphere of a restaurant. The type of operation is the focal point. The following aspects should be considered in planning.

APPEARANCE   The overall visual perception of a facility includes, among many things, exterior design, interior displays, exterior and interior lighting, and colors.

Color is the single most important part of restaurant decor and one to which customers have an immediate response. Color can be effectively used to create different types of feelings, and contrast in colors is more important than a single color itself. Primary colors (red, blue, and yellow) and secondary colors (green, orange, and violet) may be combined to produce intermediate colors. Red, orange, and yellow are considered warm colors, whereas green, blue, and violet are considered cool. Planning should be based on effective color combinations. Light has an impact on color, and each should complement the other. Certain warm colors enhance the appearance of foods. The overall concept and the food items served should therefore be considered when selecting colors. Colors that are too dark, particularly those that are reflective, should be avoided. They are tiring to the eyes of customers as well as to employees.

At restaurant operations where fast service is desired, warm colors and bright lights should be used. Cool colors are suitable for facilities located in areas with a warm climate. Light colors and bright lights are preferable in all work areas. Dark colors and insufficient light can cause employee fatigue. Blue and green are considered refreshing and are recommended for work areas.

Although a desirable, harmonious contrast in colors is preferred, care should be taken to see that there are not too many colors or designs in the same area. The atmosphere should reflect the concept in an appealing way. For room lighting as well as for color selection, emphasis should be given to focusing on objects that are designed to be highlighted, such as trademark, service mark, display cases, art objects, and decorative signs. Lighting levels and placement of lights should be such that there is no glare or strain on the eyes. If colored lights are used, they should coordinate with the colors used in the area.

ODOR   An effective exhaust system should be used for air circulation and the elimination of undesirable odors, smoke, and fumes. Good air circulation is of particular importance in restaurants that serve highly spiced or ethnic foods. Aroma is also an important factor in deciding what foods to order from the menu. The appetite-whetting aromas from broiled and baked foods, and seafood are good examples. Frequent clearing and efficient ventilation of the dining areas are desirable.

SOUND   Sound may originate from the washing of dishes, rattling of trays, silverware, or soft music played in the background. The origin of all sounds should be considered in planning the atmosphere of a restaurant. In order to avoid sounds from the kitchen and sanitation areas, sound barriers should be appropriately located. Doors opening into the dining area should not allow sound penetration each time they are opened. Silent and self-closing doors are preferred.

Music appropriate for the facility should be included as part of the atmosphere, as it has a direct effect on customer mood. Music may be used to reduce or inhibit an undesirable sound level originating in other areas as well as to enhance the atmosphere. The intensity and the quality of the sound are of particular importance. Music should cause the least possible distraction and should help portray the concept to the consumer.

COMFORT   A well-planned atmosphere should convey a feeling of comfort. A feeling of welcome and pleasure should be provided. The ideal temperature within the dining room should be between 70° and 75°F (21° to 24°C), with a relative humidity of approximately 50 percent. Temperatures vary depending on the outside climate and temperature and may be slightly lower or higher. However, dining room temperatures must be adjusted according to the type of room arrangement as well as seasonal variations.

In short, it is the combination of all these factors that results in the atmosphere desired for a restaurant concept—something that can be brought about only by careful planning. The exterior design should be coordinated with the inside decor. Parking areas, entrances, and drive-throughs should be attractive and designed to provide all possible convenience to the consumers. Landscaping obviously adds to the appeal and complements the interior atmosphere of any restaurant operation. Again, the atmosphere of a franchise restaurant should be such that it can be replicated and used in other units without undue complications and problems.

SERVICE

Service has always been a major component of a franchise restaurant. Customers have always ranked convenience and service as one of the major reasons for visiting a franchised restaurant. In fact, the expression *fast food* originated in the rapidity of the service provided by these restaurants. Any concept that will be popular depends on the type of service—table service, drive-in service, delivery service, etc. Service type selected should reflect the menu as well as the entire concept. With the advancement of technology, newer type of services are becoming popular. One example is when a customer uses an on-line computer or fax machine to place an order, which can be later picked up at the restaurant or delivered. Home delivery service using innovative methods is gaining popularity. A business in New York is serving commuters who ride on a particular train line. Before leaving the office the customers fax in their dinner orders, which are delivered when they detrain, thus providing a ready-to-eat meal package when they go home.

There is always experimentation with different concepts taking place. Some ideas survive and others die at early stages. However, proven concepts have gone through many stages, and test marketing is an essential one. Drive-through-only is rapidly gaining popularity. In one of the special franchising issues of the *Nations Restaurant News* (November 19, 1990) it was rightly predicted that the successful new concepts of the 1990s would all be based on the three C s: casualness, convenience, and comfort. Convenience to consumers should therefore be an essential component of any concept. Also shown (in Figures 11.5 and 11.6) are the menu trends and factors that have influence quick-service franchising during various stages of development.

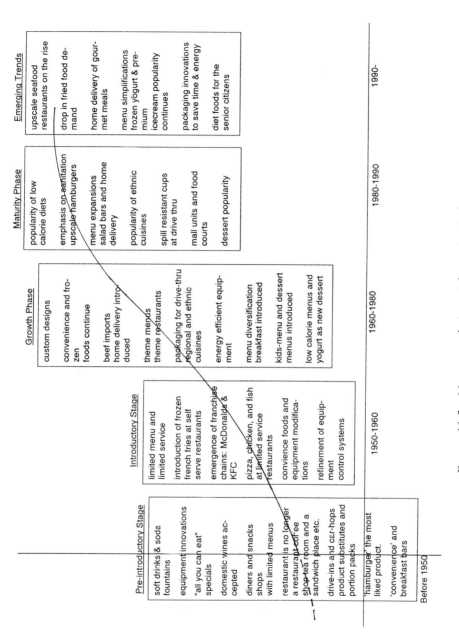

FIGURE 11.5 Menu trends in quick-service industry

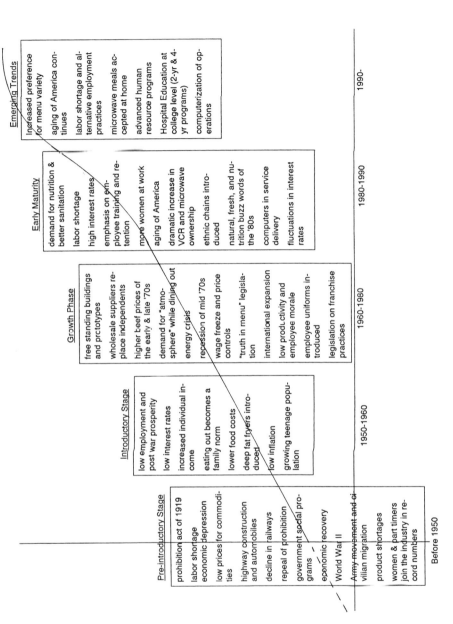

FIGURE 11.6 Contributing factors for menu trends

Marketability

The overall business concept should have a unique marketing niche that will place it a notch above the competition. This can only be confirmed by adequate consumer and field testing. Franchisors spend years before bringing out a concept in order to confirm that niche and competitive edge in the market place. A concept's life may vary and when the end approaches, modifications to that concept may be necessary. For example, hamburger chains gained peak popularity and when the competition became intense, newer menu concepts were added and are being added by many franchisors.

There should be a marked identity of the concept itself. This can be related to the product or service, such as a pasta or spaghetti restaurant or a drive-through-only concept. Another way of building marked identity is by having a trademark, service mark, logo, or other insignia that signifies the concept and that franchisees use in their businesses. Rights to such identity marks should be licensed and used in advertising and promotion whenever possible. This identity has to be carefully planned, as the business depends on the image that is perceived by the consumers.

Management

The overall suitability of the concept from the business management point of view should be evaluated. The entire concept should be transferable to other locations, nationally and internationally. An existing restaurant concept or single functioning restaurant may or may not be franchisable. Its success may be due to the location or the personality of the manager or the efficiency of the management team at a particular location. Also, concepts that are popular in certain regions may not be popular in others. These factors should be taken into consideration.

Also to be considered is the possibility of providing training and developing an operations manual. Does the system lend itself to an adequate training package and a comprehensible training manual? The business concept should run by itself after proper training. It should not be tied to the presence of a specific individual as manager. The training program in itself should be simple and comprehensive.

From the financial management point of view, the entire concept should have a built-in profit margin that should be realized at each and every location where the concept is followed. It is understood that the profit margin will vary from location to location. However, there should be enough profit margin so that the franchisee can pay the royalty fees and other fees. Franchisees are also looking for a return on their investment, so a sufficient profit margin should be provided for the success of the venture. This should be calculated before all aspects of the concept are tested and finalized.

Franchisors should also evaluate their resources before undertaking franchising. Financial resources required for the concept to take off should be assessed. A prototype of the concept must be developed. Adequate capital is needed to run the prototype operation to provide for training facilities and programs, to handle start-up costs, to provide human resources for the entire operation, and to market the concept. It is unwise to count on franchise fees as a capital resource for such expenses. The credibility of the franchisor is affected if insufficient capital resources are shown to prospective franchisees. Thus, enough capital to support the concept and its system should be available. Many innovative ideas do not fly because they lack such resources. Also to be taken into consideration is the availability of administrative, training, financial, business, legal, and computer skills necessary to develop the franchise program.

If the concept is a modification of or addition to an existing concept, as the testing expands, franchisees should be involved in evaluating the new product, equipment, and the entire system.

When all the above-mentioned attributes are tested and when the concept seems promising, it should be expanded in scale and a prototype planned. The responsibility of refinements, packaging, selection of suppliers, training, and management should be assigned to special staff of the franchisor. A package offering should be planned with consideration of all the legal aspects involved. This package should be very carefully developed by the marketing staff and should contain items such as disclosure statement, preliminary application form, menu description, service description, costs involved, fees payable, a history of the concept development, a list of the key personnel involved, and other promotional items. New concepts need especially well-planned package and marketing efforts.

Prototype Unit

A new, tested concept ready for business should be placed in pilot operation. This prototype should be conveniently located and portray the entire concept in its entirety. There can be more than one prototype unit distributed at various geographical locations. Consumer reactions to products and services can be tested at prototype locations. All attributes mentioned in this chapter should be considered in planning this prototype unit, which also becomes a model for showing to prospective franchisees. It can also be used for hands-on training of franchisees.

Factors to be considered in prototype development include site selection, design and layout, equipment selection, training, marketing, and management of the entire system. The entire operation, from receiving, storage, production, and service to quality control, should be tested and evaluated at this prototype unit. Modifications in processes, procedures, design, and management should be made based on the functioning and problems experienced at a prototype location. Quality control standards should be tested at these locations. Actual cost analyses of the entire concept can also be undertaken. All pertinent staff members of the franchisor should be involved in the development and operation of the prototype unit. Prototypes provide a good test for implementation of the operation and training programs.

Prototypes also demonstrate the interest of the franchisor in the business as well as his or her financial capability. These facilities should demonstrate genuine interest in and the feasibility of the entire business concept. Also, prototypes help in the expansion plan of the franchisors. Further strategies for development and expansion should be based on experiences gained at this prototype location. In general, in spite of the costs and effort involved, prototype operations pay off in the long run.

Restaurant Study 16

The Big Apple Bagels®

BIG IDEAS

Preopening Assistance

As a Big Apple Bagels franchise owner, you'll receive guidance and assistance in site selection, store design, and the other activities you'll need to get your new business up and running. Preopening assistance includes:

- Site selection assistance and site manual
- Lease negotiation assistance
- Store design and layout
- Guidance on inventory and equipment purchasing
- Equipment, merchandise, and services at negotiated rates
- Assistance with Grand Opening marketing campaign

Exclusive Business Components

> With the Big Apple Bagels franchise, all the systems are in place. We just need to put in the effort and enthusiasm.
>
> CHAD AND NOREEN CAPISTA, FRANCHISE OWNERS
> FLOSSMOOR, IL

Being part of the Big Apple Bagels franchise system gives you access to many powerful business tools not available to unaffiliated business owners. Exclusive business components include:

- Use of the Big Apple Bagels name and affiliated trademarks
- Use of Brewster's name and affiliated trademarks
- Use of Big Apple Bagels Confidential Operations Manual
- Access to proprietary Big Apple Bagels recipes
- Access to national buying power for food products, supplies, and services

Big Training

Intensive Training Programs

> The training I received from Big Apple Bagels was terrific. I never could have made the switch from mortgage banking to bagel baking without them.
>
> CINDY KOZELL, FRANCHISE OWNER
> WAUCONDA, IL

Our goal is to help you achieve excellence in every aspect of your business. To help you reach that ambitious goal, you and your key personnel will receive intensive training in all key areas of your operation. Our training program includes:

- Extensive classroom and in-store training
- Techniques for effectively marketing your Big Apple Bagels store
- How to hire, train, and retain good employees
- Maintaining financial controls of your store
- Preopening operational training on-site at your location

Ongoing Operational Support

> I evaluated the other bagel franchise companies and determined that the Big Apple Bagels "fresh from scratch" concept produced the best bagels in the marketplace.
>
> GEORGE SCHULZ, FRANCHISE OWNER
> OVIEDO, FL

Knowing that there's a full support team on your side, a team that understands your business and is dedicated to helping you grow, is a tremendous advantage of the Big Apple Bagels franchise program. Support components include:

- Visits from Big Apple Bagels support staff
- Periodic updates to the Big Apple Bagels Operations Manual
- Phone assistance from marketing, operations, training, and real estate experts
- Computerized cash registers that generate key sales reports
- Ongoing sales trend reports and analysis
- Quarterly newsletter filled with new ideas and procedures

I was attracted to Big Apple Bagels after learning that a number of their top people had previously owned franchises themselves. They understand the franchise owner's perspective. It's a very franchisee-driven company.

JIM AND KATHY FONTE, FRANCHISE OWNERS
MIDLAND, MI

BIG SUPPORT

Ongoing Sales-building Programs

Building unit sales is an ongoing objective for Big Apple Bagels stores. We'll provide the strategies, materials, and assistance you'll need to aggressively market your Big Apple Bagels store, including:

- Advertising and promotional strategies
- Product-specific and seasonal marketing materials
- Advertising slicks and other promotional materials
- Store-specific public relations programs
- Private-label merchandise program
- Strategies for building bulk sales and wholesale accounts

BIG APPLE BAGELS

Ongoing Program Enhancements

As a multiunit franchisee of another concept, I've been impressed with the franchisor's efforts to keep us informed and supplied with new ways to improve our operation.

TERRY AND JEAN HENDERSON, FRANCHISE OWNERS
KALAMAZOO, MI

BAB Systems, Inc. is a progressive company that knows excellence is an ongoing quest. We are dedicated to continual improvement, to searching for new products and techniques, and to becoming stronger and better each year. Our ongoing enhancement program includes:

- Ongoing research & development programs
- Access to and advice on new procedures and techniques
- Information on new vendors and improved buying programs
- Periodic refresher tr
- Access to new prod
- Franchise Advisory

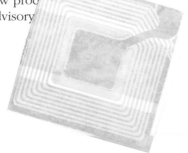

As a Big Apple Bagels franchisee, I'm really a manufacturer, retailer, and wholesaler of fresh bagels. That gives me both operating efficiencies and sales opportunities.

GARY WAPNER, FRANCHISE OWNER
YORK, PA

BIG APPLE BAGELS

The specialty coffee market is hot indeed. Between 1984 and 1994, sales of gourmet coffee grew 578 percent, making it the fastest growing segment of the $6.5 billion coffee industry. Big Apple Bagels franchisee, is a skilled *barista,* serving quality espresso, cappuccino, caffe latte, and other quality Brewster's coffees. With both fresh-brewed and whole-bean sales programs, and the natural marriage between bagels and coffee, franchisees are well-positioned to compete for a share of the local coffee market.

A natural area for expansion is wholesaling. Big Apple Bagels franchisees have the right to sell bagels to schools, convenience stores, grocery stores, restaurants, coffee shops . . . any business positioned to resell fresh, locally baked bagels.

The Big Apple Bagels franchise program is designed to help franchise owners reach their individual growth potential. In some circumstances the franchisor, BAB Systems, Inc., may allow franchisees to supplement store sales by opening satellite or kiosk, distribution-only, locations that can be supported by their store baking operation.

America's hunger for a food that's convenient, low in fat, and 100 percent delicious has turned the once-humble bagel into one of the hottest food trends in the nation. Between 1984 and 1993, yearly bagel consumption grew a whopping 169 percent. According to industry experts, bagel sales now exceed $2.5 billion yearly, and the bagel market is growing more than 20 percent annually.

Big Apple Bagels is one of the leaders of the 1990's bagel revolution. The reason for their extraordinary growth and consumer appeal commitment to excellence in all aspects of the business.

The Big Apple Bagels store operation was designed to delight not only customers, but store owners as well. A Big Apple Bagels franchise owner is able to produce freshly baked bagels in a variety of flavors and quantities, using a special system designed for simplicity and ease of operation. Unlike bagel shops that boil their bagels, Big Apple Bagels stores use a steaming method that eliminates one step in the production process, reduces equipment needs and produces a softer, more versatile bagel.

Franchise Concept Development

   There's more in store than ever before!

(a)

An interior view of a Big Apple Bagels Restaurant (courtesy BAB Holdings, Inc.)

(b)

Young or old, morning or afternoon, mealtime or snacktime, the Big Apple Bagels product line is designed to appeal to a wide range of customers at all times of the day. Each store will offers over 18 varieties of bagels, from Cinnamon Raisin and Blueberry to 8-grain to the distinctive "Everything Bagel," and just as many delicious cream cheese spreads, ranging from Chocolate Chip or Strawberry to Garden Vegetable. For customers hungry for a quick lunch or dinner, there is a range of deli sandwiches made, of course, on fresh bagels.

The Big Apple Bagels store concept and product line are specifically designed to attract bulk sales and orders to go. Fresh bagels and cream cheese are the perfect food for starting a business meeting, feeding the office, or treating the family. A Big Apple Bagels franchise owner has a wide range of aggressive promotional strategies and tools designed to sell bagels by the dozen and cream cheese by the pound.

CHAPTER *12*

Nontraditional Franchises

Nontraditional franchise restaurants are those franchises that are not the conventional freestanding restaurants. They can be double-drive-throughs, dual concepts, kiosks, stands, and can combine with other types of business. Nontraditional franchises can therefore be regarded as the wave of the new century. McDonald's, KFC, Hardee's, and all other major restaurant franchises now have some form of nontraditional franchise. McDonald's announced plans for rapid global growth with the opening of approximately 2,500 to 3,200 units in 1996, with 600 of those units in nontraditional satellite sites. This is an indication of how nontraditional restaurants will be the next wave gaining a competitive edge.

Nontraditional franchises have become a major part of the development strategy of many existing franchises. They are also referred to as *express* or *satellite* restaurants, signifying fast service.

Most of these outlets have scaled-down menus, casual counter service, lower staff costs, and limited dining area and kitchen space. Self-service products and take-out service are emphasized. Some new concepts that have emerged and are in the process of development are discussed in this chapter.

Nontraditional franchises can be classified in three categories:

1. LOCATION

Nontraditional locations for restaurants are becoming popular. These are primarily scaled-down traditional restaurants. Such locations include grocery stores, convenience stores, truck stops, amusement parks, sport arenas, service stations, hospitals, schools, universities and colleges, transportation facilities, highway travel plazas, shopping centers, and retail outlets.

Dual-concept and *co-branded franchising* refer to franchising of two different concepts and brands at one location. Co-branding or dual branding is successful where there is a good synergy between two or more concepts, such as breakfast and lunch business, cold and hot food items, burgers and Mexican foods, and fish and chicken items. A common example is the food/fuel dual concept. McDonald's Corporation signed deals with petroleum companies to develop food and fuel locations. It entered into alliance with Amoco Corporation to develop co-branded locations throughout the Midwest and Northeast and the mid-Atlantic, north central, and southeastern states. This agreement followed a similar deal with Chevron Corporation. KFC has a dual-branding arrangement with sibling brand Taco Bell. Sales have increased due to this alliance because it provides a variety of choices to consumers.

Co-branding is also being practised with other concepts. Carlson Hospitality Worldwide has launched triple-branded complexes that fuse its Country Inn & Suites, Italianni's, and Country Kitchen concepts under one roof. Del Taco, Inc., teamed up with Mrs. Winner's

Chicken and Biscuits to offer dual-branded business. Blimpie International, Baskin-Robbins, and Dunkin' Donuts have entered into a co-branding test agreement. Several burger chains are negotiating or for already have dual branding with such brands as Long John Silver's, Lee's Famous Recipe Chicken, and Captain D's. Flagstar's grilled chicken, El Pollo Loco, forged a dual-branding alliance with Fosters Freeze, the soft-serve dessert serving chain. Blimpie, a submarine sandwich chain, is co-branding with Uni-Mart, Inc. Other major franchisors are moving in the same direction.

These types of new alliances are vehicles to expand market share, especially as petroleum companies have monopolies in prime locations. The point of distribution for both businesses is increased because the concept goes where the consumers are. Such alliances are also being made with retail stores, convenience stores, truck stops, educational institutions, stadiums, airports, hospitals and theme parks.

2. TYPE OF RESTAURANT

Another way of classifying nontraditional restaurants is based on the type of restaurant, such as deli, bakery, kiosk, stand, coffee shop, and delivery service only.

Coffee Shops

Gourmet coffee shops are popping up on every corner. The demand for coffee continues to increase. Based on the rapid development, all indications are that there will be more than 10,000 coffee shops by the year 2000. Coffee shops offer multiple varieties of coffee in several styles of restaurant. For example, Coffee Beanery® offers four options for franchising its upscale retail coffee shops: espresso carts, on-line stores, kiosks, and storefront cafes. Each has a different investment level with varying menu options, from muffins to scones to deli sandwiches.

Gourmet coffee's popularity is based on the special coffee beans used. There are two types of coffee beans, robusta and arabica. Robusta beans commonly found in grocery stores, are grown in lower altitudes, and have less body, texture, and flavor than arabica beans, which are grown at higher altitudes and have greater body, texture, flavor, and aroma. Even though the taste of arabica beans is far richer than robusta beans, they contain less than half the caffeine. Gourmet coffees come in a variety of flavors. For example, Gloria Jean's Gourmet Coffees® has seventy-five varieties of gourmet coffee. In addition to coffee, it sells gourmet imported and domestic teas, coffee grinders, coffee and espresso machines, coffee and tea accessories, fine china and porcelain products, and exclusive packaged gift sets. Freshly brewed coffee, espresso, espresso-based drinks, and iced coffee drinks are also offered by many of their stores. Gourmet coffee shops are located in high-traffic areas and may range from full stores to kiosks or carts. Java Coast® franchisors of fine gourmet coffees have stores in conjunction with I Can't Believe It's Yogurt®, thus providing hot and cold foods under the same roof.

Pretzels

Due to the increasing health food consciousness of consumers, pretzels, particularly the soft ones, are becoming popular. Their sales figures keep on growing. Several types of restaurant sell these low-fat items. For example, Auntie Anne's began the company with a single store in a Pennsylvania farmer's market. Since then, the soft pretzel chain has grown to more than three hundred franchised units. Soft pretzels come in flavors ranging from original to garlic to raisin combined with several choices of dip.

Bagels

Bagels are also gaining popularity and fulfilling the need of a growing market, most of them being health-conscious consumers. Per capita bagel consumption in the past ten years have increased close to 113 percent. Bagels are offered in multiple varieties with or without other food items. For example, Manhattan Bagel Company offers eighteen varieties of New York–style bagels, twenty-five kinds of cream cheese, and various sandwiches and beverage choices. They have also entered into an agreement with Texaco Gas Stations and Star Mart convenience stores for co-branding. For Bruegger's Bagel Bakery®, hearth-baked bagels are the specialty item. Bruegger's operates more than three hundred stores and plans to have more than a thousand by 2000.

Take-out and Delivered Foods

Although pizza is one of the most commonly delivered food items, other nontraditional foods are also offered as take-out or delivered foods. According to the National Restaurant Association, 50 percent of restaurant orders are for off-premise consumption. A new concept, Takeout Taxi® delivers restaurant-prepared meals to the increasing number of homes without time to cook. Such third-party-delivered items are becoming popular.

3. TYPE OF MENUS

Nontraditional menu items can form another basis for classifying restaurants. Examples include restaurants that serve specialized or limited menus such as water bars, juice bars, tea and coffee shops, ice cream and yogurt shops, potato bars, and vegetarian food outlets.

Although it is hard to draw a distinct line at this time, in later years, with the development of businesses and menu items, there will be clear lines of differentiation among them.

REASONS FOR GROWTH

Looking at the spurt in growth of nontraditional franchises, it is easy to predict that freestanding restaurants will become almost obsolete or at least will face tough competition. Alternative development usually means dealing with a partner, which can be from any of the facilities listed earlier. The reasons for the rapid development of nontraditional franchises are given below:

- The main idea behind nontraditional franchises is very simple. Instead of building a freestanding restaurant where consumers come in, the food is taken to where consumers are in large numbers or visit frequently. For example, people visit shopping malls frequently, much of the time with family and children, and that provides a good opportunity for serving foods that can be conveniently purchased and easily consumed. In other words, there is a captured clientele and already existing customer base.
- One of the major problems facing the restaurant industry is the lack and costliness of space for expansion. Many franchises are located at prime sites, which are very expensive. Nontraditional franchises provides less expensive space because there is sharing of facilities. Also, as these concepts do not require large square footage, they are easily affordable when compared to free-standing restaurants.
- Consumer demand is primarily responsible for the growth of nontraditional franchises. Consumers are becoming more electronic or computer oriented. More and more people are spending time in front of their PC or TV watching videos or playing video games. Such activities lend themselves to picking up foods from a convenient location. For example, after grocery shopping, one can select videos and also pick up food at the same location.
- The development of food courts, super retail stores, expanded grocery shops, multipurpose student centers, and new travel plazas all lend themselves favorably for the development of nontraditional franchises. Consumers benefit from the convenience of getting more than one product or type of goods in a single stop.

- Technological developments have allowed entrepreneurs to introduce scaled-down versions of restaurants with computerized equipment, providing express facilities. This is evident from the growing number of kiosks or units in airports, theme parks, and gas stations. Cooking equipment that uses the latest technology facilitates the preparation of menu items in a short time. Some foods can be pre-prepared and stored with limited reheating or preparation before service. For example, a few years back it took almost half an hour or more to cook one pizza. With the development of technology that can keep the frozen dough as well as ovens that can cook rapidly, large number of pizzas can be prepared in a very short period. Such equipment includes ventless fryers, electronically controlled ovens, pressure cookers, and infrared heaters. Also, technology has facilitated purchase using credit cards without even getting out of the car. For example, one can purchase gas at a service station and simultaneously order food on the intercom located at the pump, pay by credit card, and get the food from the drive-through or the counter. This convenience and captured clientele will definitely lead to the growth and development of nontraditional franchises.
- With the changing status of the economy and demographic conditions, more and more people are working away from home. With limited lunch time, it becomes much easier for customers to buy food from a convenient location. On the flip side of this argument is the fact that with the development of computer technology, many large companies are letting employees work at home. This again emphasizes the need to bring prepared food from a convenient location or have it delivered.

ADVANTAGES OF NONTRADITIONAL FRANCHISES

1. The building of the restaurant is quicker and easier because there is an existing facility. There is also less deal-

ing with zoning permits, curb-cut requirements, parking requirements, etc., as these are already been taken care of by the host facility.
2. As discussed earlier, there are already existing customers whose primary purpose is to buy other goods, such as in retail stores and gas stations. In some retail stores and stadiums, the large number of customers provide a captured clientele that finds the availability of foods convenient. Particularly in shopping centers, where families with children are present, nontraditional franchises provide a place for rest as well as a convenient location for purchasing food. Also, employees working in large retail stores and shopping centers themselves will be good customers. Having a co-branded store, particularly with a well-known franchisor, can bring in more customers to the host operation. For example, having a McDonald's or Taco Bell store at a gas station may increase the number of customers coming to the gas station.
3. Because there is always a host, nontraditional franchises piggyback on many fundamentals such as space, utilities, insurance, and infrastructure. The limited space requirement leads to considerable savings.
4. There can be shared labor in some places where host employees can help in simultaneously working for the restaurant operations. For example, employees at a convenience store may also help in selling food for the restaurant. Similarly, cleaning, trash pick-up and rest room maintenance can be done by shared employees.
5. Safety is considered one of the major benefits of the nontraditional franchise. Restaurants in the midst of other retail establishments with lighted facilities and large numbers of people are less likely to be targets of crime. Also, the cameras and security staff/system in some shopping centers and stores provide added security at practically no extra cost. Traditional restaurants that have experienced problems due

to insecurity are finding nontraditional locations attractive for development.

6. From the franchisee's point of view, the franchise fee for nontraditional franchises is much lower than for others. Also, the initial investment is drastically reduced. Various incentive programs are provided by franchisors to franchisees to develop these outlets. Some franchisors are waiving initial fees and charging reduced royalties. Other incentives are provided for expansion of similar units. Nontraditional franchises are also easier to expand by developing other outlets.

7. Nontraditional franchises are cost-effective. Because their equipment and space needs are limited, costs required to purchase, utilize, and maintain them are reasonable. In the case of dual concepts, this advantage increases considerably because the facilities and equipment can be shared.

8. It is easier to get financing because the amount and risk involved are considerably less for nontraditional franchises. Some financial institution have issued loans to cover the entire cost of start-ups. This is a very uncommon practice in the case of freestanding restaurants.

9. Advertising and promotion costs are drastically reduced, as there is a customer base already visiting the location. Also, by combining promotional activities with the host, considerable savings in marketing can be achieved.

10. Rental costs or financing for building a restaurant prohibits many interested parties from investing in such a venture. Leasing the space from another entity reduces the costs incurred in purchasing or securing financing.

11. The management of restaurants is much easier with the reduced staff and operational activities; thus, labor and operational costs are greatly reduced. Some operational aspects, as for cleaning and trash disposal, are reduced due to the sharing of expenses. For dual concepts, cooperative buying can also help reduce the costs.

12. Innovative merchandising techniques can be used, such as displaying food preparation in glass peek-in windows, venting the aroma of baked goods to attract consumers, providing food samples to the consumers, flashing electronic messages, providing specials hours when items are discounted, and using a public address system for advertising.
13. Because many of the nontraditional franchises do not provide seating, costs for purchasing, cleanup, and maintenance are considerably reduced. Also, there is no liability issue if seating is not provided by the restaurant.
14. Specialized menu items designed for carrying out and that can be easily prepared can be developed. This lends to innovation and creativity, which, many franchisees miss in traditional operations. These items can reflect local tastes and flavors.
15. Parking facilities, handicap access facilities, and the need to provide rest rooms are major expenses for traditional restaurants that are practically eliminated when a host has already taken care of them.
16. Consumers who are loyal to a brand can conveniently find their favorite food without having to drive to the location for the sole purpose of buying food. They also get attracted to the brand name when they find it while they are in the midst of shopping or conducting other business activities.

Disadvantages of Nontraditional Franchises

Although there are many advantages of nontraditional franchises, there are certain drawbacks that should be considered, which are listed below:

1. Because of the scaled-down operation, there are limited choices of menu items available in these restaurants. This

eventually cuts down on the margin of profits available; therefore, a large volume of business is necessary. Also, limited equipment and space hinders productivity. This becomes a problem when large quantities of food must be prepared in a short period of time.

2. Most of the nontraditional franchises are located in heavily traveled areas. Due to space and labor constraints, the service is not as fast as one would expect from an express location. This has led to long waiting lines with uncomfortable customers, particularly during peak periods of time. Also, the cleanliness of the counters and trash cans may be affected by labor constraints.

3. Due to the lack of sufficient equipment, space, and labor, the quality of food may be adversely affected. It becomes difficult to control time/temperature parameters, which affects the quality of food as well as its safety.

4. Franchisees of traditional restaurants may find themselves at a disadvantage if they encounter unfair competition from within by new nontraditional franchises close to their location. Some franchisees who are in business for a number of years have found that smaller units have moved in across the street in a stadium or student center, thereby cutting into their market share. Some franchisees are discontented because the old franchise agreements did not refer to nontraditional franchises. Some of the operators of gas stations and convenience stores may want to enter as franchisees, thereby increasing fears among existing franchisees. Nontraditional locations can cannibalize sales of traditional restaurants within the same franchise. This could put a strain in the franchisor-franchisee relationship.

5. Because of the symbiotic relationship with the host, much depends on the health of its business. If traffic diminishes, the restaurant also suffers. The concept is based on continued traffic to the host business. No one is interested in com-

ing to a kiosk in a shopping center just for the dining experience. The nontraditional franchise is only as good as the host site.
6. Franchisors may be judged by the company they keep. The decision pertaining to the selection of a location has to be taken very carefully. A store with a poor image or performance may have direct impact on the restaurant business. Also, the concept of the nontraditional franchise should lend itself to the host business. Is it appropriate to sell pastries at a service station surrounded by oil stains, engine fumes, and noisy auto repairs?
7. One of the advantages mentioned earlier is the sharing of labor that can take place at host site. However, this can also affect the quality of food and service. How many people would like to be served food by someone who has just dispensed gasoline at the full service station? How much is one likely to appreciate getting a taco prepared by someone who was on the cash register of a convenience store? Again, there is a risk of sacrificing the quality of food and service.
8. Customers expect the same level of service they associate with the brand name of a franchise, whether the venue is traditional or nontraditional. These expectations are affected by any adverse experience like cold hamburgers, soggy tacos, lack of cleanliness, or a long wait in line. These perceptions may have a lasting impact because the customer does not differentiate a traditional from a nontraditional franchise.
9. The above-mentioned adverse aspects may quickly erode the image created by hard work of the franchisor. The lack of dining experience, seating, and supplies may turn off loyal consumers.
10. Due to the lack of storage facilities, the likelihood of running out of supplies and food is great. Such shortages result in not having items desired by the consumer or send-

ing employees who are needed for operations to other restaurant to fetch the items in short supply.
11. The contract with the host must be carefully examined. An increase in lease payments or short-term termination can affect the restaurant business. Also, there should be limitations on signing up a competing restaurant at the same location.
12. There are chances of increased competition where more than one branded or nonbranded restaurant is present at the same location. This can result in thin distribution of the customer base, as seen in food courts and student centers. The competition can exist even with the host. For example, at some gas stations, beverages are major income providers. If restaurants enter the same location, there may be competition for this income-generating menu item.

Restaurant Study 17

McDonald's

McDonald' is the world's largest quick-service restaurant organization, operating more than 11,800 family restaurants in the United States and in 114 international markets. More than 22 million people visit McDonald's each day. Approximately 75 percent of McDonald's restaurant businesses worldwide are owned and operated by independent, local businesspeople; the remainder are operated by the company. In June 1967, McDonald's first international restaurant opened in Richmond, British Columbia, Canada. During the 1950 and 1960s, McDonald's restaurants were traditionally located in suburban areas. Today, restaurants are also located in small towns, urban centers, shopping malls, tollways, hospitals, airports, military bases, and college campuses, among other sites.

All McDonald's restaurants operate according to a strict quality assurance system. McDonald's operating motto is "Q. S. C. & V.," which stands for Quality, Service, Cleanliness, and Value. While McDonald's original menu focused on the hamburger, today it had grown to include chicken, fish, salads, and a full breakfast menu, including Egg and Sausage McMuffin sandwiches and fresh-baked buttermilk biscuit sandwiches. Other products include McDonald's world famous french fries, and McDonaldland cookies, milkshakes, and sundaes.

Through the years, McDonald's has developed innovative food service technology, setting the standards for its industry.

In 1961, McDonald's founded Hamburger University (H.U.), its first formal training center, in Elk Grove Village, Illinois. Today it is an international management training center for McDonald's franchisees, restaurant managers, and management personnel. In 1983, the University moved to its new home on a 130,000-square foot facility in Oak Brook, Illinois. The first H.U. located outside of the United States opened in Tokyo in 1972; Munich followed in 1975 and London in 1982.

The first prototype McDonald's restaurant was a red-and-white tile building with two golden arches, the symbol which McDonald's subsequently adopted for use as the company logo. With the steady expansion of business and the trend among customers away from the practice of eating in their cars, McDonald's began experimenting with larger buildings and inside seating. The first restaurant with inside seating opened in

Huntsville, Alabama, in July 1966. In 1969, McDonald's introduced a completely new restaurant design, featuring a brick building with mansard roof, large expanses of windows, and inside seating for approximately 120 customers. The first drive-thru was built in 1975. The current architectural developments at McDonald's are the additions of drive-thrus and playland facilities. Restaurants with different themes, reflected in the interior design, have also also built in recent years.

Seventy-five percent of McDonald's restaurant businesses are owned and operated by independent, local businesspeople. It is estimated by McDonald's that on an average day, 22 million people around the world visit McDonald's. The first floating McDonald's restaurant is located on a Mississippi River boat anchored near the St. Louis Arch. McDonald's has opened restaurants in such nontraditional locations as St. Joseph's Hospital in Phoenix, Arizona; Penn Station in Newark, New Jersey; a Las Vegas casino; an automobile showroom in Japan; and a race course in Hong Kong. Among many McDonald's restaurants located in unique sites are several in buildings dating back to the early 13th century. These are located in Shrewsbury, England, in a building that dates back to 1220; and in Frieburg, West Germany. It recently opened a restaurant in Soviet Union.

Some of McDonald's fascinating McFacts are listed below:

- If lined up end to end, McDonald's 80 billion hamburgers served would go from earth to the moon and back over 19 times.
- McDonald's prepares more than 2 million pounds french fries every day to meet customers' demands.
- The northernmost McDonald's is located on the Arctic Circle in Finland. The southernmost McDonald's is located in Christchurch, New Zealand.
- McDonald's cracks open more than *3 million* fresh Grade A eggs every morning to serve breakfast to its customers worldwide.

RAY A. KROC, FOUNDER, MCDONALD'S CORPORATION

At age 52, Ray Kroc was the exclusive distributor for a company that produced Multimixer milkshake machines. He was impressed by a small chain of hamburger restaurants in San Bernardino, California. He acquired franchising rights from the owners, the McDonald brothers. He then founded McDonald's Corporation in 1955. He bought out the McDonald brothers for $2.7 million, borrowed at interest rates that eventually made the cost $14 million. The deal turned out to be a bargain, since

McDonald's grew into the world's largest restaurant organization with 11,800 restaurants in the United States and 53 other countries. Mr. Kroc's dedication to strict standards—his diligence to providing customers with consistent quality, service, cleanliness, and value—and his innovative use of cooking techniques made him the pioneer in this type of restaurant business. He created an enterprise made up of thousands of small businesses, run by independent franchisees who own and operate 75 percent of McDonald's restaurants. "If you work just for money, you'll never make it," Mr. Kroc often said, "but if you love what you're doing and you always put the customer first, success will be yours."

Through the years, Ray Kroc was involved in charitable activities involving the welfare of children, diabetes, arthritis, multiple sclerosis, and chemical dependency. He served as chairman of McDonald's Corporation from its founding in 1955 until 1977, when he was named senior chairman of the board. Kroc was the recipient of many prestigious awards, including the Horatio Alger award, and he was inducted posthumously into the Advertising Hall of Fame in 1989. Ray Kroc died on January 14, 1984, in San Diego, California.

MILESTONES

1948   Dick and Mac McDonald open the first McDonald's drive-in restaurant in December in San Bernardino, California.

1954   Ray A. Kroc, a Multimixer salesman from Oak Park, Illinois, visits Dick and Mac's San Bernardino McDonald's. His curiosity is initially aroused by the large number of Multimixers they were buying. Ray Kroc became *franchising agent* for the McDonald brothers.

1955   April 15, Ray Kroc's first McDonald's opens in Des Plaines, Illinois. Opening day sales were $366.12. It rained.

1957   McDonald's formally adopts its QSC (Quality, Service, and Cleanliness) operating motto.

1958   McDonald's sells its 100 millionth hamburger.

1961   Hamburger University opens in the basement of the Elk Grove Village, Illinois, McDonald's restaurant and conferred bachelor of hamburgerology degrees to the first graduating class. The McDonald brothers sell the company to Ray Kroc.

1963   McDonald's serves its one billionth hamburger. Ronald McDonald makes his debut in Washington, D.C. Filet-O-Fish sandwich becomes the first new menu item since the original menu.

1963	McDonald's offers its common stock to the public at $22.50 per share. The following year, the stock is listed on the New York Stock Exchange.
1966	Ronald McDonald makes his first national television appearance in Macy's Thanksgiving Day Parade.
1967	McDonald's first restaurant outside the United States opens in Richmond, British Columbia, Canada. Indoor seating is introduced.
1968	The Big Mac and Hot Apple Pie are added to the menu.
1969	McDonald's serves its five billionth hamburger. International Division is formed.
1972	Introduction of large fries.
1973	Quarter Pounder added to the national menu.
1973	The Egg McMuffin breakfast sandwich invented. First McDonald's in a college facility opens at the University of Cincinnati.
1974	The first Ronald McDonald House opens in Philadelphia to serve as a "home away from home" for families of children being treated at nearby hospitals. McDonald's opens its first operation in a zoo at Toronto, Canada. McDonaldland Cookies introduced.
1975	The first McDonald's drive-thru opens in Sierra Vista, Arizona, offering customers quick service without leaving their cars. McDonald's celebrates its 20th anniversary.
1976	The 20 billionth hamburger served.
1977	McDonald's officially adds a complete breakfast line to its national menu.
1979	The 30 billionth hamburger served.
1980	The first floating McDonald's opens on the Mississippi Riverfront in St. Louis, near the famous Jefferson Memorial Arch. McDonald's celebrates its 25th (silver) anniversary.
1983	Chicken McNuggets introduced in all U.S. McDonald's.
1984	The 50 billionth hamburger served in New York City. Ray Kroc, founder and senior chairman of the board of McDonald's dies January 14.
1985	Ray Kroc's first McDonald's restaurant restored to its original form and reopens in May as the McDonald's museum in Des Plaines, Illinois.
1987	Fresh, ready-to-eat salads introduced in all U.S. McDonald's. McDonald's serves its 65 billionth hamburger.

1988 McDonald's serves its 70 billionth hamburger.
1989 Ground-breaking takes place on the first McDonald's in Moscow. The restaurant has five dining rooms and seats 700 people.

McDonald's History Listing

1948 In December, Dick and Mac McDonald open the first McDonald's drive-thru restaurant in San Bernardino, California. A little hamburger man called "Speedee" becomes the company symbol.
1954 Ray A. Kroc, a Multimixer salesman from Oak Park, Illinois, visits Dick and Mac's San Bernardino McDonald's, his curiosity initially aroused by the large number of Multimixers they were buying.
Ray Kroc becomes franchising agent for the McDonald's brothers.
1955 On April 15, Ray Kroc opens his first McDonald's in Des Plaines, Illinois.
In July, Ray Kroc opens his second McDonald's in Fresno, California.
Total sales for the company are $193,772.
1956 McDonald's Corporation adds 12 restaurants including Chicago, Skokie, Waukegan, Joliet, and Urbana, Illinois; Hammond, Indiana; Los Angeles, Torrence, and Reseda, California; and Dallas, Texas.
Ray Kroc hires Fred Turner as a grillman in his #1 store in Des Plaines.
1957 McDonald's becomes known for the motto "QSC" for Quality, Service, and Cleanliness.
Ray Kroc personally delivers free hamburgers to Salvation Army workers in Chicago at Christmas.
At year-end, sales for McDonald's 40 restaurants totaled $3,841,327.
1958 McDonald's sells its 100 millionth hamburger.
Fred Turner becomes Vice President of the company.
McDonald's annual sales skyrocket 151 percent over the previous year to $10,896,163.
1959 100th restaurant opens in Fond du Lac, Wisconsin.
In total, a record 66 restaurants open.
McDonald's begins billboard advertising.

1960	"Look for the Golden Arches," was McDonald's first jingle. The company celebrates its 5th anniversary, opening its 200th restaurant in Knoxville, Tennessee, and selling its 400 millionth hamburger. Annual sales total $37,579,828. Ad campaign cheers on the "All American Meal"—a hamburger, fries, and milkshake.
1961	Ray Kroc buys out McDonald brothers for $2.7 million. Hamburger University opens in basement of the Elk–Grove Village, Illinois, McDonald's restaurant and confers Bachelor of Hamburgerology degrees to the first graduating class. The U.S. Secretary of Agriculture, Charles Murphy, eats the 500 millionth McDonald's hamburger. The Golden Arches—a modernistic M—replaces "Speedee" as the company's logo.
1962	"Go for Goodness at McDonald's," a new advertising logo slogan is introduced. McDonald's sells its 700 millionth hamburger.
1963	The 500th McDonald's restaurant opens in Toledo, Ohio. The one billionth McDonald's hamburger is served by Ray Kroc on the Art Linkletter Show. Ronald McDonald makes his debut in Washington, D.C. The Filet-O-Fish sandwich becomes the first new menu addition since the original menu. Lou Groen, McDonald's franchisee, creates the sandwich.
1964	At year-end, there are 657 restaurants. The company's gross sales hit $130 million.
1965	McDonald's celebrates its 10th anniversary with the first public stock offering at $22.50 per share. Average annual sales for a McDonald's restaurant are $249,000. Ronald McDonald makes his first appearance in the Macy's Thanksgiving Day Parade. Television network advertising begins.
1966	Ronald McDonald appears in his first national television commercial. On July 5, McDonald's is listed on the New York Stock Exchange with the ticker symbol MCD. McDonald's exceeds $200 million in sales, and sells its two billionth hamburger. McDonald's in Huntsville, Alabama, becomes the first restaurant with inside seating.

1967	McDonald's All-American High School band is organized. The first international McDonald's restaurants open in Canada and Puerto Rico.
1968	The Big Mac and Hot Apple Pie are added to the menu. The 1,000th store opens in Des Plaines, Illinois. McDonald's opens in Hawaii. Average annual sales for McDonald's restaurants open at least 13 months are $33,000.
1969	International Division is formed. McDonald's serves 3.5 million hamburgers per day. The "billions served" sign changes to "five billion." The new McDonald's building design is introduced to replace the "red-and-white" design.
1970	Christmas gift certificates introduced. The 1,500th restaurant opens in Concord, New Hampshire. A McDonald's in Bloomington, Minnesota, is the first to reach $1 million in annual sales. McDonaldland becomes the setting of a new series of commercials created for children.
1971	"You Deserve a Break Today—So Get Up and Get Away to McDonald's" becomes the new advertising theme. Hamburglar, Grimace, Mayor McCheese, Captain Crook, and the Professor join Ronald McDonald in McDonaldland. Home Office moved from Chicago to Oak Brook, Illinois. McDonald's opens in Japan, Germany, Australia, Guam, Holland, and Panama. The first McDonald's Playland opens in Chula Vista, California.
1972	The 10th and 11 billionth hamburgers are sold. McDonald's becomes a billion-dollar corporation on December 17. The 2,000th store opens in Des Plaines, Illinois. Ray Kroc receives the Horatio Alger Award. Large fries introduced.
1973	2,500th store opens in Hickory Hills, Illinois. Quarter Pounder added to the national menu. First McDonald's opens in a college faculty at the University of Cincinnati.
1974	3,000th McDonald's opens in Woolwich, England. McDonald's opens its first restaurant in a zoo in Toronto, Canada.

Fred Turner becomes President and Chief Executive Officer.
The first Ronald McDonald House opens in Philadelphia, Pennsylvania.
The company sells its 15 billionth hamburger.
By year-end, total sales for the company approach $2 billion.
McDonald's Cookies introduced.
"Twoallbeefpattiesspecialsaucelettucecheesepicklesonionsonasesameseedbun" promotion introduces our most famous advertising jingle for the Big Mac.

1975 Egg McMuffin added to the national menu.
The first drive-thru is established in Sierra Vista, Arizona.
McDonald's celebrates its 20th anniversary.
The new campaign, "We Do It All For You," is introduced in April.

1976 The 4,000th store opens in Montreal, Canada.
McDonald's sales surpass $3 billion.
The 20 billionth hamburger is sold.
"You, You're the One" advertising campaign introduced.

1977 A complete breakfast line is added to our national menu.
The first McDonald's All American High School Basketball Team is organized.
Ray Kroc celebrates his 75th birthday.
McDonald's announces plans to secure the land (80 acres) for its home office in Oak Brook, Illinois.

1978 The 5,000th restaurant opens in Kanagawa, Japan.
The 25 billionth hamburger is served.
The 15,000th student graduates from Hamburger University.
Filming store for McDonald's commercials opens in the City of Industry, California.

1979 McDonald's introduces its new advertising theme "Nobody Can Do It Like McDonald's Can."
The 30 billionth hamburger is sold.
Happy Meals are added to the national menu.
The average annual sales for a restaurant in existence for more than 13 months exceeds $1 million for the first time.
Four years after the domestic drive-thru concept was introduced, nearly half (2,884) of McDonald's restaurants have such a facility.

1980 McDonald's celebrates its silver (25th) anniversary.
Birdie, the Early Bird, joins the McDonald's characters.

	The first floating McDonald's is launched on the historic Mississippi riverfront in St. Louis, just South of the famous Gateway Arch.
	The 6,000th restaurant opens in Munich, Germany.
1981	McDonald's renews its most successful campaign ever—"You Deserve a Break Today."
	There are 5,554 stores in the United States and 1,185 internationally, totaling 6,739 stores worldwide with sales of $6 billion.
	The first Ronald McDonald House outside the United States opens in Toronto, Canada by the end of 1981, Ronald McDonald Houses provided shelter annually for an estimated 33,000 families.
	First McDonald's restaurants in Spain, Denmark, and the Philippines open.
	Total Playlands systemwide at year-end were 979.
1983	McDonald's restaurants are now located in 32 countries around the world.
	The new Hamburger University opens in Oak Brook, Illinois.
	McDonald's opens its 7,000th restaurant in Falls Church, Virginia.
	There are 7,778 restaurants at year-end.
	Chicken McNuggets introduced into all domestic restaurants by year-end.
	McDonald's serves its 45 billionth hamburger.
1984	Ray Kroc, Founder and Senior Chairman of the Board of McDonald's, dies on January 14.
	McDonald's restaurants open in four new countries: Andorra, Finland, Taiwan, and Wales.
	Chicken McNuggets introduced in Canada, Japan, France, and Germany, making McDonald's the second largest purveyor of chicken in the world.
	In June, McDonald's introduces a new national advertising theme, "It's a Good Time for the Great Taste of McDonald's."
	The 8,000th McDonald's restaurant opens in Duluth, Georgia.
	McDonald's serves its 50 billionth hamburger. Dick McDonald, who along with his brother Mac, started the McDonald's system 41 years earlier, ate the ceremonial burger.
	McDonald's year-end systemwide sales surpass $10 billion.

McDonald's celebrates the 10th anniversary of the Ronald McDonald House with a national fund-raiser that generates more than $5 million. At year-end, there are 73 Ronald Houses in the United States, Canada, and Australia.

In 1984, a new McDonald's restaurant opens somewhere in the world every 17 hours. (A new store has opened every 18 hours for the past 10 years.)

The average sales volume for a McDonald's restaurant is $1,264,000.

1985 McDonald's adds seven new countries to its roster: Thailand, Luxembourg, Bermuda, Venezuela, Italy, Mexico, and Aruba. At year-end, McDonald's restaurants are located in 43 countries worldwide.

The first European Ronald McDonald House opens in Amsterdam, Holland. By December, there are more than 90 Ronald McDonald Houses open worldwide.

On April 15, McDonald's celebrates its 30th year of operation. On April 20, McDonald's celebrates its 20th anniversary as a publicly owned company.

Ray Kroc's first McDonald's restaurant is restored to its original form and reopens in May as the McDonald's #1 Store Museum in Des Plaines, Illinois.

In August, McDonald's "Large Fries for Small Fries" promotion featuring Mary Lou Retton, helps raise more than $2.6 million for the Muscular Dystrophy Association.

McBlimp, the world's largest airship, appears in the skies over New York City, introducing McDonald's newest form of advertising.

McDonald's serves its 55 billionth hamburger.

Millions sing "The Hot Stay Hot and the Cool Stays Cool" as McDonald's introduces the McDLT sandwich on November 4.

December tops off McDonald's most successful year to date with systemwide sales of more than $11 billion.

1986 McDonald's serves one out of every four breakfasts eaten outside of the home in the United States.

In March, McDonald's adds fresh baked buttermilk biscuit sandwiches to the breakfast menu. McDonald's employees bake more than 1.5 million buttermilk biscuits daily.

McDonald's breaks the "sound barrier" by producing "Silent Persuasion," the first-ever television commercial featuring sign language and closed captioning for the hearing-impaired audience.

The Golden Arches greet customers for the first time in Argentina, Cuba, and Turkey.

McDonald's restaurants worldwide change their road signs to read "More Than 60 Billion Served."

The Long Island Jewish Medical Center hosts the opening of the 100th Ronald McDonald House in September.

McDonald's becomes the first fast-food restaurant to provide the public with a complete food product ingredient listing.

On December 1, McDonald's opens the first fast-food restaurant in North Pole, Alaska. The restaurant is located on Santa Clause Lane.

Year-end systemwide sales top $12 billion.

1987 Freshly tossed salads are added to the McDonald's national menu on May 15.

McDonald's grants Sears Roebuck & Co. the rights to carry a children's line of clothing called "McKids."

McDonald's serves its 65 billionth hamburger.

1988 *Fortune* magazine names McDonald's hamburgers among the 100 products America makes best.

George Cohon, President and CEO McDonald's Restaurants of Canada Limited, signs an agreement with the Soviet Union to open a McDonald's in Moscow, with a possible 19 more restaurants to follow.

McDonald's adds three countries to its roster: Hungary, Yugoslavia, and Korea.

McDonald's opens its 10,000th restaurant in Dale City, Virginia, on April 6.

COSMc joins the other characters in McDonaldland.

McDonald's serves its 70 billionth hamburger.

Year-end systemwide sales top $16 billion.

1989 McDonald's welcomes its 2,000th franchisee in the United States.

In May, all U.S. restaurants begin serving the country-style McChicken sandwich.

The Big Mac celebrates its 21st birthday.

McDonald's opens its 11,000th restaurant in Hong Kong on October 20.

McDonald's Restaurants of Canada Limited brings its restaurant expertise to the Toronto Sky Dome. This first-of-a-kind

	stadium location has 4 restaurants, 20 Skysnack locations, 48 beverage stations, and a team of stadium vendors. McDonald's serves its 75 billionth hamburger. Year-end systemwide sales top $17 million.
1990	On January 31, the first McDonald's restaurant in Moscow opens. McDonald's, Ronald McDonald Children's Charities and Worldwide Fund join forces to produce a new environmental education booklet for kids entitled "WEcology." On April 1, no-fat, no-cholesterol apple bran muffins and whole grain cereal are added as permanent breakfast items in all of McDonald's U.S. restaurants. McDonald's celebrates its 35th birthday on April 15. On April 17, McDonald's announces "McRecycle USA," an environmental program that sets a goal of at least $100 million annually to purchase recycled materials. On April 21, the largest audience in U.S. history ever to watch a Saturday morning entertainment program views "Cartoon All-Stars to the Rescue," an anti-substance abuse special. In May, McDonald's introduces a new, national advertising theme, "Food, Folks, and Fun." The worldwide total of Ronald McDonald Houses reaches 138. The first McDonald's restaurant constructed primarily out of recycled materials opens in Albany, New York. In July, all McDonald's U.S. restaurants post complete McDonald's food product nutrition and ingredient information. On July 18, McDonal's announces a decision to cook french fries in 100 percent cholesterol-free vegetable oil. McDonald's and the Environmental Defense Fund form a task force to find new ways to reduce, reuse, recycle, and compost solid wastes produced by our restaurants worldwide. The first McDonald's restaurant opens in Shenzhen, a special economic zone of the People's Republic of China. McDonald's serves its 80 billionth hamburger. Year-end systemwide sales surpass $18 billion. A total of 641 restaurants are added in 1990, including 335 outside of the United States. On November 19, McDonald's enters its 54th country with the opening of a restaurant in Chile.

1991	McDonald's opens its 12,000th restaurant on March 22 in New Hyde Park, New York.
	McDonald's introduces the McLean Deluxe sandwich which features a 91 percent fat-free beef patty.
	On November 12, McDonald's enters its 57th country with the opening of a restaurant in Greece.
	Hamburger University celebrates its 30th anniversary.
	McDonald's nationally phases in recycled carry-out bags and recycled fiber napkins.
	Leaps & Bounds, a new indoor family play center, opens in September in Naperville, Illinois.
	The 150th Ronald McDonald House opens in Paris, France.
	On October 31, United Airlines begins offering McDonald's Friendly Skies Meals to kids.
1992	Year-end sales top $21.8 billion.
	On July 7, McDonald's serves its 90 billionth hamburgers.
	McDonald's food is now being served in two restaurant cars on the Swiss Federal Railroad. This marks McDonald's first train operation.
	New advertising theme, "What You Want Is What You Get" debuts.
	McDonald's receives the 1991 "Green Thumbs Up" award in recognition of our environmental education and beautification efforts.
	The world's largest McDonald's opens in Beijing, China. This two-story 28,000 square foot facility seats more than 700 and employs 1,000.
	Baked apple pie replaces the fried apple pie on the standard menu.
	McDonald's opens in "Six Flags" theme park in St. Louis, Missouri.
	National Association of Secondary School Principals recognizes McDonald's for exemplary leadership in education.
	McDonald's opens in Warsaw, Poland, breaking records of opening day sales.
1993	McDonald's opens its doors inside a Wal*Mart store. The restaurant is approximately 1,250 square feet, and includes a limited menu of hamburgers, cheeseburgers, Quarter Pounders, french fries, Coca-Cola and diet Coke, and sausage biscuits.

In February, McDonald's opens its 13,000th restaurant in Acapulco, Mexico. This represents the 60th McDonald's operating in Mexico since opening in this country in 1985. It features 210 seats within its three levels, employs more than 90 crew members, and includes an "Auto-mac," Mexico's term for a drive-thru.

The first McDonald's at sea opens March 14 aboard the *Silja Europe*, the world's largest ferry, which transports vacationers across the Baltic Sea between Stockholm and Helsinki.

In April, McDonald's marks the milestone of serving more than 95 billion hamburgers.

Big Mac celebrates its 25th anniversary.

McDonald's opens its second restaurant in Moscow—the Ogareva McDonald's.

On September 3, McDonald's opens in its 67th country—Iceland.

On October 14, McDonald's opens in its 68th country—Tel Aviv, Israel. The 450 seat Canyon Avalon restaurant serves our traditional, nonkosher menu, with kosher meal.

1994 Annual sales reach $25.9 billion.

Grand openings of Ronald McDonald Houses in Rio de Janeiro, Brazil, and Auckland, New Zealand, bring the total number of houses to 163 in 12 countries.

For the first time, five Middle Eastern countries enjoy the Golden Arches with restaurant openings in Egypt, Oman, Kuwait, Bahrain, and the United Arab Emirates.

McDonald's opens in Bulgaria, Trinidad, New Caledonia, and Latvia, bringing the total number of countries McDonald's does business in to 79.

1995 McDonald's celebrates its 40th anniversary.

The Fajita Chicken Salad is introduced nationally.

The new advertising theme "Have You Had Your Break Today?" debuts.

McDonald's opens its doors to Estonia, Romania, Malta, Columbia, Jamaica, Slovakia, South Africa, Qatar, St. Maarten, and Honduras, bringing the total number of countries to 89.

McDonald's and The Walt Disney Company sponsor The American Teachers Awards honoring 36 teachers from across the United States.

1996   The Arch Deluxe introduced in May and the Deluxe Line is launched in September.
        McDonald's introduces a new Grilled Chicken Salad Deluxe and new Garden Salad along with Caesar and Fat-Free Herb Vinaigrette dressing in November.
        McDonald's acquires 184 Roy Rogers restaurants in the Baltimore and Washington, D.C., areas.
        The first McSKI-THRU opens in Lindvallen, Sweden, on December 7.
        McDonald's opens in Croatia, Western Samoa, Fiji, Liechtenstein, Lithuania, India, Peru, Jordan, Paraguay, Dominican Republic, Belarus, and Tahiti, bringing the total number of countries to 101.
        McDonald's signs a ten-year contract with The Walt Disney Company. McDonald's will become Disney's primary promotional partner in the restaurant industry, sharing exclusive marketing rights in more than 100 countries, and linking McDonald's restaurants to Disney theatrical releases, theme parks and home video releases.
1997   McDonald's U.S.A. reorganizes into five divisions, providing more focused attention on serving our customers.
        "Did Somebody Say McDonald's?" is introduced as the new advertising theme.
1998   McDonald's opened at Disney Marketplace in Orlando, Florida.

INTERNATIONAL OPENINGS

Countries Opened	Date Opened
1. United States	04/15/55
2. Canada	06/01/67
3. Puerto Rico	11/10/67
4. Virgin Islands	09/04/70
5. Costa Rica	12/28/70
6. Guam	06/10/71
7. Japan	07/20/71
8. Netherlands	08/21/71
9. Panama	09/01/71
10. Germany	11/22/71
11. Australia	12/30/71
12. France	06/30/72

	Countries Opened	Date Opened
13.	El Salvador	07/20/72
14.	Sweden	11/05/73
15.	Guatemala	06/19/74
16.	Netherland Antilles	08/16/74
17.	England	10/01/74
18.	Hong Kong	01/08/75
19.	Bahamas	08/04/75
20.	New Zealand	06/07/76
21.	Switzerland	10/20/76
22.	Ireland	05/09/77
23.	Austria	07/21/77
24.	Belgium	03/21/78
25.	Brazil	02/13/79
26.	Singapore	10/20/79
27.	Spain	03/10/81
28.	Denmark	04/15/81
29.	Philippines	09/27/81
30.	Malaysia	04/29/82
31.	Norway	11/18/83
32.	Taiwan	01/28/84
33.	Andorra	06/29/84
34.	Wales	12/03/84
35.	Finland	12/14/84
36.	Thailand	02/23/85
37.	Aruba	04/04/85
38.	Luxembourg	07/17/85
39.	Bermuda (closed)	07/24/85
40.	Venezuela	08/31/85
41.	Italy	10/15/85
42.	Mexico	10/29/85
43.	Cuba	04/24/86
44.	Turkey	10/24/86
45.	Argentina	11/24/86
46.	Macau	04/11/87
47.	Scotland	11/23/87
48.	Yugoslavia	03/24/88
49.	Korea	03/29/88
50.	Hungary	04/30/88
51.	Russia	01/31/90
52.	People's Republic of China	10/08/90

Countries Opened	Date Opened
53. Chile	11/19/90
54. Indonesia	02/23/91
55. Portugal	05/23/91
56. Northern Ireland	10/14/91
57. Greece	11/12/91
58. Uruguay	11/18/91
59. Martinique	12/16/91
60. Czech Republic	03/20/92
61. Guadeloupe	04/08/92
62. Poland	06/17/92
63. Monaco	11/20/92

Photos showing different concepts of McDonald's restaurants (a) at an Oasis in Vinita, Oklahoma; (b) at the Denver International Airport; (c) at 30th Street Train Station in Philadelphia, Pennsylvania; and (d) at a Navy Pier in Chicago, Illinois (photos courtesy of McDonald's Corporation)

Nontraditional Franchises

Countries Opened	Date Opened
64. Brunei	12/1292
65. Morocco	12/18/92
66. Saipan	03/18/93
67. Iceland	09/03/93
68. Israel	10/14/93
69. Slovenia	12/02/93
70. Saudi Arabia	12/08/93
71. Oman	07/30/94
72. Kuwait	06/15/94
73. New Caledonia	07/26/94

(e)   (f)   (g)   (h)

Photos showing different concepts of McDonald's restaurants (e) at a gas station in Beebee, Arkansas; (f) at the Old Orchard mall in Skokie, Illinois; (g) in Wal*mart Store in Baton Rouge, Louisiana; and (h) at Family Entertainment Center in Little Rock, Arkansas (photos courtesy of McDonald's Corporation)

INTERNATIONAL OPENINGS *(continued)*

Countries Opened	Date Opened
74. Egypt	10/20/94
75. Trinidad	11/12/94
76. Bulgaria	12/10/94
77. Bahrain	12/15/94
78. Latvia	12/15/94
79. United Arab Emirates	12/21/94
80. Estonia	04/29/95
81. Romania	06/15/95
82. Malta	07/07/95
83. Colombia	07/14/95
84. Jamaica	09/28/95
85. Slovakia	10/13/95
86. South Africa	11/11/95
87. Qatar	12/13/95
88. St. Maarten	12/15/95
89. Honduras	12/14/95
90. Croatia	02/02/96
91. Western Samoa	03/02/96
92. Fiji	05/01/96
93. Liechtenstein	05/03/96
94. Lithuania	05/31/96
95. India	10/13/96
96. Peru	10/18/96
97. Jordan	11/07/96
98. Paraguay	11/21/96
99. Dominican Republic	11/30/96
100. Belarus	12/10/96
101. Tahiti	12/10/96
102. Ukraine	05/28/97
103. Cypress	06/12/97
104. Macedonia	09/06/97
105. Ecuador	10/09/97
106. Bolivia	10/23/97
107. Reunion Island	12/14/97
108. Isle of Man	12/15/97
109. Suriname	12/18/97

INDEX

A & W Restaurants, Inc., 11, 52–59
　milestones, 55–59
Acknowledgment, 104
　required by FTC Rule, 104
ADI, *see* Areas of dominant influence
Advertisement, 69, 71–72
　fees, 215
　fund, 215
　promotion, 69, 71–72
Advertising manual, 172
Agreement, 89–113
　franchise, 108–113
　franchising, 108–113
Application, 146
　forms, 152–157, 158–163
　process, 146–166
Arby's, 114–117
　history, 114–115
　philosophy, 117
Areas of dominant influence (ADI), 185
Assets, 230
　current, 230
　fixed, 232
Average restaurant check, 235

Balance sheet, 230–232
　current assets, 230
　current liabilities, 232
　fixed assets, 232
　fixed liabilities, 232
　net worth, 232
Bankruptcy, 93
　history, 93
Basic Disclosure Document, 92
Baskin-Robbins, 11, 363
Bennigans, 24
Beverage cost percentage, 235
Big Apple Bagels, 353-359

Blimpie International, 265–268, 363
Bruegger's Bagel Bakery, 365
Burger King, Corp., 206–212
　historical facts, 208–212
Business expansion, 75
Business-format franchising, 7–8
　examples, 7

Captain D's, 174–176, 363
Chicken, 23, 48–51
　restaurants, 48–51
　segment, 48–51
Chuck E. Cheese Pizza, Inc., 48
Churchs Chicken, 138–140
　history, 138–140
Churchs Fried Chicken, Inc., 24
Coco's, 24
Code of Ethics, 151
Coffee Beanery, 363
Coffee shops, 363
　Coffee Beanery, 363
　Gloria Jean's Gourmet Coffees, 364
　Java Coast, 364
Colonel Harlan Sanders, 11
Community involvement support, 201
Company-owned franchises, 101
Concept, 5, 315
　established, 315
　franchise, 315–352
Costs, 218–219
　building, 224
　equipment, 218, 224
　food, 227
　labor, 227
　opening, 218
　preopening, 222
　real estate, 224
　sitework, 224

393

Costs (cont'd)
  training, 218
Cost control, 226–227
  in restaurant operations, 226–227
Country Kitchen, Inc., 362

Dairy Queen, Inc., 12, 237–241
Del Taco, Inc., 362
Development fee, 216
Disclosure, 13, 89, 92
  document, 92, 94
    contents, 94–113
Domino's Pizza, Inc., 309–313
Dual-concept franchising, 9
Dunkin' Donuts, Inc., 12, 363

Earnings claim, 106–107
El Pollo Loco, 363
Equipment selection and layout, 339–346

FAC, see Franchisee Advisory Committee
FTC, see Federal Trade Commission
Federal Trade Commission (FTC), 8
Fee, 5, 214
  franchise, 5, 214, 216
  development, 216
  renewal, 217
  royalty, 215, 216
  standard, 216
  training, 218
  transfer, 217
  variable, 216
Field support manual, 172
Financial control assistance, 200
Financial information, 103
  franchisors, 103
Financing, 99
  arrangement, 99
Flagstar's Grilled Chicken, 363
Food cost, 227
Food cost percentage, 227, 235
Franchise, 172, 89–113
  advertising manual, 172
  agreement, 89–113
  application, 146
  application forms, 152–157, 158–163
  application process, 146–166
  arrangements, 6
  cancellation, 100

company-owned, 101
concept, 5, 315–352
concept development, 315
definition, 4
Disclosure Act, 13
fee, 5, 214
field support manual,
financial information, 103
marketing manual, 172
mass advertising, 69, 71–72
number of, 101
offering circular, 92
operations manual, 166–170
operations package, 166
prototype unit, 351–352
public figure involvement, 103
quality control manual, 173
renewals, 73, 100
restaurants, 172
rule, 89, 107
site analysis, 185–190
site inspection manual, 173
site selection, 102
system performance, 74–75
termination, 73, 100
training manual, 170–171
training programs, 102
transfers, 73
Franchise agreement, 108–113
  termination, 73
  transferability, 73
Franchise arrangements, 6
  types, 6
Franchise concept development, 5, 315
  features of successful concepts, 315
  menu development, 316–322
Franchise Disclosure Act, 13
Franchised restaurants, 42–51
  chicken, 23, 48–51
  evaluation, 133-135
  hamburger, 23, 42–45
  Mexican, 23
  operations manual, 166–170
  operations package, 166
  pizza, 23
  restaurant operator's training, 166–170
  sandwich, 23, 42–45
Franchisee, 64–70
  advantages of franchising to, 64–70

contribution, 76–77
definition, 4
disadvantages of franchising to, 64, 70–75
franchise selection by, 124–131
preopening training, 193
qualifications, 120–124
recruitment, 79
retention, 79
selection, 79, 119–120, 131–137
selection procedure, 124–131
self-evaluation, 131–133
training, 191–196
training programs, 171–172, 195
Franchisee-franchisor relationship, 243-264
friction points, 243—264
relationship considerations, 243-264
Franchise fee, 5, 214
Franchise Restaurant Segments, 42–51
chicken, 23, 48–51
hamburger, 23, 42–45
Mexican, 23
pizza, 23, 45–48
sandwich, 42–45
seafood, 23
steak, 23
Franchise rule, 107
violation of, 107
Franchise types, 7–9
business-format, 7–8
conversion franchising , 9
dual-concept franchising, 9
master franchisor program, 9
nontraditional franchises, 361–373
product-and-trade-name, 6–7
Franchising, 64
advantages, 64
agreement, 108–113
basic concepts, 4
business-format franchising, 7–8
challenges of, 136–137
co-branded franchising, 362
conversion, 9
costs, 218–219
definition, 2–4
disadvantages, 64
dual-concept franchising, 362
in the economy, 21–51
global, 304–308

history and development, 10–14
international, 269
introduction, 1
legal documents, 92, 94
master, 295
product-and-trade-name, 6–7
pros and cons, 63-80
regulations, 92, 94
symbiotic relationship, 5
Franchising documents, 92, 94
evaluation, 92
Franchisor, 64–78
advantages of franchising to, 64, 75–78
bankruptcy history, 95–96
business experience, 95
concerns, 243
definition, 4
disadvantages of franchising to, 64, 78–80
evaluation, 135
field services,
litigation history, 95
master, 9
obligation, 202–203
before opening, 202
ongoing, 202–203
qualifications, 124–131
research and development, 200–201
selection, 124–131
service provided by, 181–203
training programs, 171–172, 195
Franchisor-franchisee relationship, 243-264
franchisee concerns, 245–255
franchisor concerns, 255–258
friction points, 244–245
relationship considerations, 258–261
relationship regulations, 261–264
FTC Rule, 8, 89

Gloria Jean's Gourmet Coffees, 364
Godfather's Pizza, Inc., 85–87, 146
Golden Corral, Inc., 60–61
Gross sales, 215

Hardee's Food Systems, Inc., 147–149, 361
application forms, 158–163
nontraditional franchises, 361

History and development of franchising, 10–14
  early 1900s, 11
  future growth, 13
  in 1950s, 11–12
  in 1960s, 12
  in 1970s, 12–13
  in 1980s, 13
  in 1990s, 13–14
Howard Johnson Restaurant, 11

Initial investment, 219
IFA, see International Franchise Association
International Dairy Queen, Inc., 237–241
International Franchise Association (IFA), 3
  code of ethics, 151
International franchising, 269
  Africa and Middle East, 307
  assessment of environmental factors, 300–304
  decision-making model, 298–300
  direct, 293
  European Common Market, 306–307
  global market, 304–308
    Africa and Middle East, 307
    European Common Market, 306–307
    North and South America, 308
    Pan-Pacific Region, 305–306
  international expansion, 271–278
    factors related to, 271–278
  international markets, 271
  joint ventures, 296–298
  master franchisees, 295
  methods of, 293-298
  North and South America, 308
  Pan-Pacific Region, 305–306
  points to consider in, 278–293
International House of Pancakes, 12
Investment, 219
  initial, 219
Italianni's, 362

Java Coast, 364
Joint ventures, 296–298

KFC, see Kentucky Fried Chicken Corp.
Kentucky Fried Chicken Corp., 5, 23, 361, 362

Colonel Harlan Sanders, 11
  nontraditional franchises, 361
Kroc, Ray, 11, 375–376

Labeling, 322–323
  nutritional, 322–323
Labor cost, 227
Labor cost percentage, 227
Layout and physical facilities, 332
  dining, 336–338
  equipment selection and layout, 332
  preparation area, 335–336
  receiving areas, 332–333
  serving, 336–338
  storage areas, 333-335
    dry, 333
    frozen, 334–335
    refrigerated, 334
Lee's Famous Recipe Chicken, 363
Legal documents, 89–113,
  franchising, 108–113
License agreement, 89–113
Liabilities, 232
  current, 232
  fixed, 232
Long John Silver's Seafood Shoppes, 81–84, 363
  history, 81–84

McDonald's Corporation, 23, 24, 316, 361, 362, 374–392
  application form, 152–157
  Hamburger University, 131, 192, 374
  history, 378–388
  international openings, 269, 388–392
  Krock, Ray, 11, 375–376
  milestones, 376–378
  nontraditional franchises, 361
  site selection, 183
  training program, 192
Made in Japan Teriyaki Experience, 204–205
  history, 204
Manhattan Bagel Company, 365
Manuals, 166–173
  advertising, 172
  field support, 172
  marketing, 172
  operating, 166–170

Index 397

quality control, 173
site inspection, 173
training, 170–171
Market analysis, 185–191
 access, 189
 area characteristics, 188
 availability of services, 190
 competition, 190
 cost consideration, 188–189
 market, 190
 physical characteristics, 188
 position, 189
 traffic information, 189–190
 type of restaurant and service, 190
 utilities, 189
 visibility, 190
 zoning, 185–188
Marketing, 185–191
 fund, 216
 manual, 172
 support, 196–198
Master franchisee, 295
Master franchisor, 295
 program, 295
Materials management, 198–199
Menu development, 316–322
 food characteristics, 318
 nutritional quality, 322
Mexican, 23
 restaurants, 23
 segment, 23

National Sanitation Foundation, 339
Nontraditional franchises, 361–373
 advantages, 367–370
 classification, 362
 definition, 361
 disadvantages, 370–373
 express restaurants, 361
 satellite restaurants, 361
NSF, see National Sanitation Foundation
Nutritional labeling, 322–323
Nutritional quality, 322–328

Occupational Safety and Health Act, (OSHA), 196
Operating manual, 166–170
 contents, 166–170
Operating package, 166

Operational support and field service, 199–200
Orange Julius, 12, 241
OSHA, see Occupational Safety and Health Act

Pizza, 23
 restaurant, 23
 segment, 23
Pizza Hut, Inc., 5, 23, 24
Pizzeria Uno, 141–144
 history, 141
Product-and-trade-name franchising, 6–7
Profit and loss statement, 227–230
 administrative and general expenses, 229
 advertisement and promotion expenses, 229
 controllable expenses, 229
 depreciation, 230
 direct operating expenses, 229
 employee benefits, 229
 gross profit, 229
 music and entertainment expenses, 229
 net profit, 230
 payroll, 229
 profit before depreciation, 230
 profit before income tax, 230
 profit before rent, 230
 rent or occupation costs, 230
 repairs and maintenance expenses, 229
 total controllable expenses, 230
 total income, 229
 utilities expenses, 229
Projections, 220, 221
 sales, 220
Promotion, 69
Proprietary marks,
Prototype unit, 351–352

Q.S.C. & V., see Quality, Service, Cleanliness, and Value
Quality, Service, Cleanliness, and Value (Q.S.C.&V.), 375
Quality control, 66
 manual, 173

Ratio analyses, 232–236
 accounts receivable to total revenue ratio, 232, 233

Ratio analyses *(cont'd)*
  accounts receivable to turnover ratio, 233, 234
  acid test ratio, 232, 233
  activity ratio, 234
  average restaurant check, 233
  beverage cost percentage, 235
  beverage inventory turnover ratio, 234
  current ratio, 232, 233
  debt-equity ratio, 234
  debt-to-total-assets ratio, 234
  fixed asset turnover ratio, 234
  food cost percentage, 227, 235
  food inventory turnover ratio, 234
  liquidity ratio, 232
  management proficiency ratio, 236
  net profit to net return ratio, 236
  net return on assets ratio, 235
  number of times interest earned ratio, 234
  operating efficiency ratio, 235
  operating ratio, 235, 236
  profitability and rate of return ratio, 234
  profit margin ratio, 235
  quick ratio, 232
  return on assets ratio, 235
  return on owner's equity, 235
  solvency, 233
  solvency ratio, 234
Recipe standardization, 328–332
  standardization procedures, 329–332
  standardized recipes, 328
Recurring funds, 97
Red Lobster, 5
Research and development (R & D), 200–201
Restaurant design, 191
  factors to be considered in, 191
Restaurant study, 11, 15–20, 23-24, 52–59, 60–61, 81–84, 85–87, 114–117, 138–140, 141–144, 147–149, 174–176, 177–180, 204–205, 206–212, 237–241, 309–313, 353-359, 361–362, 374–392
  A & W Restaurants, Inc., 11, 52–59
    milestones, 55–59
  Arby's, 114–117
    history, 114–115
    philosophy, 117
  Big Apple Bagels, 353 359
  Burger King Corporation, 206–212
    historical facts, 208–212
  Captain D's, 174–176, 363
    history, 174
  Church's Chicken, 138–140
    history, 138–140
  Dairy Queen, Inc., 12, 237–241
  Domino's Pizza, 23, 309–313
  Godfather's Pizza, 85–87, 146
  Golden Corral, 60–61
  Hardee's, 23, 147–149
    application form, 158–163
    site selection, 183
  International Dairy Queen, Inc., 237–241
  Kentucky Fried Chicken, 5, 361, 362
  Long John Silver's Seafood Shoppes, 81–84
    history, 81–84
  McDonald's, 23, 374–392
    application form, 152–157
    Hamburger University, 192
    site selection, 183
    training program, 192
  Made in Japan Teriyaki Experience, 204–205
    history, 204
  Pizza Hut, 5, 23
  Pizzeria Uno, 141–144
    history, 141
  Red Lobster, 5
  Subway, 23, 177–180
  Wendy's International, Inc., 15–20, 23
    historical highlights, 17–20
Risk, 67
  minimum, 67
  reduction, 67
Royalty fee, 215, 216
Ruby Tuesday, 24

Sales, 215
  estimated, 220
  gross, 215
  per seat, 235
  projection, 220
Sanitation area, 338–339

Selection, 119
  franchisee, 119–120
  site, 185
Service mark, 4
Services, 72, 181–203, 346–351
  mark, 4
  provided by franchisor, 72
Site analysis, 185–190
  access, 189
  area characteristics, 188
  availability of services, 190
  competition, 190
  cost consideration, 188–189
  market, 190
  physical characteristics, 188
  position, 189
  traffic information, 189–190
  type of restaurant and service, 190
  utilities, 189
  visibility, 190
  zoning, 185–188
Site and market analysis, 185–191
  access, 189
  area characteristics, 188
  availability of services, 190
  competition, 190
  cost consideration, 188–189
  market, 190
  physical characteristics, 188
  position, 189
  traffic information, 189–190
  type of restaurant and service, 190
  utilities, 189
  visibility, 190
  zoning, 185–188
Site inspection manual, 173
Site selection, 183-190
Standardized recipes, 328–332
  standardization procedures, 328–332
Standards, 66–67
Subway, 23, 177–180

Taco Bell, Inc., 23, 362
Take-out and delivered foods, 365

Takeout Taxi, 365
Tastee-Freeze, 12
Termination, 73
  franchise, 73
  franchise agreement, 89–113
  renewals, 73
Trademark, 4, 89, 94
Trade name, 4
Training, 191–196
  advantages of, 192–193
  Churchs chicken's, 194
  crew member, 195
  formal, 171
  hands-on, 171
  initial opening, 195
  KFC's, 194
  manual, 170–171
  McDonald's, 192
  ongoing, 171, 195
  potential franchisee training, 193
  preopening, 193
  programs, 171–172, 195
    types of, 193
  restaurant operator's, 195
  types of, 193
Transfers, 73
  franchise, 73
Tricon Global Restaurants, 5
Types of franchising, 6–9
  business-format, 7–8
  conversion franchising, 9
  creative franchising,
  dual-concept franchising, 9
  of franchise arrangements, 6
  master franchisor program, 9
  product-and-trade-name franchising, 6–7

UFOC, see Uniform Franchise Offering Circular,
Uniform Franchise Offering Circular, 93-94
  contents, 94–113

Wendy's International, Inc., 15–20